T0299227

A Systematic Guide to Leadership Selection Using Total Quality Management Techniques

The old cliché states that not every manager is a leader, but the more important part of that sentiment is that to be a *good* manager, one *has to* be a good leader. This perception is because *good* managers do more than manage. They have to lead by inspiration, they have to lead by example, and they have to lead through the best times for their organizations as well as the absolute worst times.

A Systematic Guide to Leadership Selection Using Total Quality Management Techniques identifies the application gap and presents a methodology based on Total Quality Management (TQM) to support the guidance of a process to select leadership (at any level of the organization). A modification to the House of Quality and a product of the Massachusetts Institute of Technology is presented and discussed as the core of a leader selection process. Two case studies are used to reinforce the concepts and applications. Finally, the book introduces some experiments for leadership development using virtual worlds and ends with a note for the future using the metaverse and digital twins for leadership.

The book is intended for professionals and executives wanting to learn more about leader selection, engineering and business students, directors of human resources, and researchers in the field of leadership.

A Systematic Guide to Leadership Selection Using Total Quality Management Techniques

Luis Rabelo
Charles W. Davis, Jr.
Ahmed Elattar
Hamed M. Almalki

CRC Press
Taylor & Francis Group
Boca Raton London New York

CRC Press is an imprint of the
Taylor & Francis Group, an **informa** business

First edition published 2023

by CRC Press
6000 Broken Sound Parkway NW, Suite 300, Boca Raton, FL 33487-2742

and by CRC Press
4 Park Square, Milton Park, Abingdon, Oxon, OX14 4RN

CRC Press is an imprint of Taylor & Francis Group, LLC

Library of Congress Cataloging-in-Publication Data
Names: Rabelo Mendizabal, Luis C. (Luis Carlos), 1960- author. | Davis, Charles W., author. | Elattar, Ahmed, author.
Title: A systematic guide to leadership selection using total quality management techniques / Luis Rabelo, Charles W. Davis, Jr., Ahmed Elattar, Hamed M. Almalki.
Description: 1 Edition. | Boca Raton, FL : CRC Press, [2023] | Includes bibliographical references.
Identifiers: LCCN 2022046433 (print) | LCCN 2022046434 (ebook) | ISBN 9781032342474 (hardback) | ISBN 9781032342498 (paperback) | ISBN 9781003321170 (ebook)
Subjects: LCSH: Leadership. | Total quality management.
Classification: LCC HD57.7 .R313 2023 (print) | LCC HD57.7 (ebook) | DDC 658.4/092—dc23/eng/20221026
LC record available at https://lccn.loc.gov/2022046433
LC ebook record available at https://lccn.loc.gov/2022046434

ISBN: 978-1-032-34247-4 (hbk)
ISBN: 978-1-032-34249-8 (pbk)
ISBN: 978-1-003-32117-0 (ebk)

DOI: 10.1201/9781003321170

Typeset in Times
by codeMantra

Dr. Rabelo would like to dedicate this book to his two beloved sons, Luis Gisli and Carlos Magnus.

Dr. Ahmed Elattar would like to dedicate this book to his loving wife Nerine, and three children, Lana, Liam, and Raya.

Dr. Charles W. Davis, Jr. would like to dedicate this book to his son, the late Brian S. Thomas, and his late nephew, Albert J. Davis.

Dr. Hamed M. Almalki would like to dedicate this book to his loving children, Eilaf, Jana, Muhammad, Ibrahim, Abdurrahman and Tamim.

Contents

Preface

This book presents an innovative method of selecting a leader based on systems engineering concepts and the voice of the customer. We introduce this methodology from project teams to chief executive officers. A crucial decision to make is who will lead. The purpose of the book is to illustrate the ubiquitous importance of leadership.

In this book, readers will understand that there are several issues to consider when making such a decision, for example, issues regarding what responsibilities the team leader will have, what the goals of the team/organization are, and match that to the qualifications that the candidate should have to take on a leadership role. It is always important to have the input of the stakeholders in this decision which is beyond a beauty contest of a set of Curriculum Vitae.

The book's design allows for learning leadership style classification, the Matrix of Change, and how the principles of Total Quality Management should guide the process. When selecting a team leader, you need to look at what their responsibilities as a leader will be. Will the leader oversee meetings, decide who will work on what, determine how the work will be done, handle conflict resolution among team members, keep account of performance and keep managers apprised of the team's performance and progress? A very important part is the current and desired situation of the team or organization in which the leader will be appointed.

The book is intended for professionals and executives wanting to learn more about leader selection, engineering and business students, directors of human resources, and researchers in the field of leadership. The content of the book introduces the following fundamental concepts:

- Leadership Challenges
- Leadership Classification
- Total Quality Management (TQM) and the Matrix of Change
- Case Study – Yahoo to exercise the methodology
- Advances in how to develop leadership skills using Virtual Reality/Metaverse

In small, medium, and large organizations, leadership is recognized by all as one of the most important virtues. Effective leaders understand how to empower their teams and lead their organizations to greatness, which means "building" leaders who can help a business achieve lasting success. Unfortunately, true leaders are not always easy to identify, so recruiters sometimes accidentally place followers or leaders who emphasize a type of leadership not desired by the context or situation into positions that require effective leaders. In addition, different types of leaderships are adequate for different contexts and situations. Therefore, when recruiting for a leadership role, recruiters/decision-makers/stakeholders can apply the guide to help them find the true leader amongst their applicants.

Foreword

By Luis Rabelo

Several events impacted my decision to pursue research in leadership. First, in graduate school, I attended classes in engineering management and administration, where the methodology of the case study (written reports and discussions in class) was used. Unfortunately, after several weeks, the discussions were forgotten, and there were no clear systematic ways to synthesize leadership. Second, I come from an engineering background. While at MIT Sloan (as a student of a Joint MS Degree in Engineering and Management), I had the opportunity to attend a course taught by Dr. Erik Brynjolfsson in Electronic Commerce in the late 1990s. I saw firsthand the power of the Matrix of Change (which I easily related to systems engineering). Third, I attended the Executive Education certificate program in "The Art and Practice of Leadership Development" at Harvard University, which Dr. Ron Heifetz led in the late 2010s. With Dr. Ron Heifetz, I experienced and observed a new way to teach leadership beyond just the traditional case study. Several parts of this book are a consequence of those three events. Fourth, while working at NASA as a project manager, I visited the NASA Kennedy Space Center Library. While I went through several books during my tenure, one book will always stand out. The book title was "How NASA Builds Teams" by Charles Pellerin. This book changes how its readers approach team-building exercises and new and foreign projects. It helps you shift the way you analyze certain problems and the way you can come up with a systematic methodology to build and guide a highly effective team. I will never forget how much I enjoyed reading it and how much it inspired me to start thinking about problems more systematically.

Acknowledgments

We want to acknowledge all the individuals and organizations that contributed in one way or another to this endeavor. First, we would like to express our sincere thanks to Ms. Rana Riad and Dr. Sayli Bhide for their efforts in virtual modeling. Those were our first steps in virtual worlds. We appreciate Dr. Wilawan Onkham and Dr. Muyuan Li for their help with the Matrix of Change in the case of Apple Maps. Our gratitude to the late Dr. Barbara Truman, who was very instrumental in assisting as an advisor for immersive learning and her help with collaboration with the UCF's Institute for Simulation and Training. Thanks to Dr. Debra Hollister for providing subject matter experts in Psychology and developing scenarios to design the virtual worlds for Dr. Almalki's and Dr. Davis' research work. We thank Ms. Gwenette Writer Sinclair for designing, developing, and optimizing 3D virtual worlds and avatars with the OpenSim software platform utilizing Dreamland Metaverse, a leading OpenSim software host. Finally, of course, to the SDM Program at MIT and MIT Sloan to let one of our authors attend the course of Professor Erik Brynjolfsson in Electronic Commerce, where he learned the Matrix of Change when he was an MS student at MIT. Last but not least, our thanks to Taylor & Francis for publishing this book and for their infinite patience. We are very positive that this book is a great contribution to the area of leadership and a unique one in the challenge of leader selection.

Author Biographies

Luis Rabelo, PhD, was the NASA EPSCoR Agency Project Manager and is currently a Professor in the Department of IEMS at the University of Central Florida. He received a joint degree in Electrical and Mechanical Engineering from Panama and master's degrees from the Florida Institute of Technology (1987) and the University of Missouri (1988). He received a PhD in Engineering Management from the University of Missouri in 1990, where he also did post-doctoral work in Nuclear Engineering in 1991. In addition, he holds a dual MS degree in Systems Engineering & Management from the Massachusetts Institute of Technology. He has contributed to over 300 publications. He holds 3 international patents, and he has published 3 books and graduated 50 doctoral students as advisor/co-advisor.

Dr. Ahmed Elattar is a Product Manager working in the software and IT industry and is also an Adjunct Professor in the field of Systems Engineering. He received his bachelor's degree in Systems Engineering with a minor in Computer Science in 2005 and received his master's degree in Engineering Management in 2010, both of which were from The George Washington University. After spending a few years working in the manufacturing space, he completed his doctorate in 2014 at the University of Central Florida in Industrial Engineering. His dissertation was focused on Change Management and modeling/simulating executive transitions and succession planning.

Ahmed has experience in manufacturing, systems architecture, project management, and Information Technology. In the past, he has held various positions with Symantec Corporation, EGAT Group, and ECG-NACO International. His career today is focused on Customer Engagement, Technical Marketing Engineering, and driving user insight into the Research and Development process.

Hamed M. Almalki, PhD initially worked at Saudi Airlines as an Avionics and Reliability Engineer. After that, he worked as Change Management Consultant at Saudia Aerospace Engineering Industries (SAEI). With the same company, he served as talent acquisition developer and technical leadership instructor. Recently Dr. Almalki joined Taif University as an

Assistant Professor in the Industrial Engineering Department. He received his Electrical Engineering degree from the American University of Sharjah, master's degree in business administration from the Rollins College, and his Master's and PhD in Industrial Engineering from the University of Central Florida. His major research contribution is in the field of Engineering Leadership.

Charles W. Davis, Jr., PhD, is Professor at Valencia College teaching various courses in Pre-Engineering, Mathematics, Quality Assurance and Test Methods, Engineering Management, Communications, and Ethics and Statistical Theory for Engineering Technology. He has over 27 years of extensive experience in both public and private sectors consisting of Project Management, Leadership, and Quality Assurance. He earned a bachelor's degree in design engineering, a master's degree in Industrial Engineering, and a PhD in Industrial Engineering with a specialization in Engineering Management focusing on Undergraduate Leadership from the University of Central Florida. In terms of professional accomplishments and community involvement, he has authored and published two articles and co-authored two published articles. He is also a member of the Florida Engineering Society, National Society of Black Engineers, and Tau Beta Pi Engineering Honor Society.

Investigating the Leader Selection Process

1

For all organizations (i.e., government, academic, for-profit/non-profit), there comes a time when a change must take place in leadership. It consumes much thought and planning to ensure the right decision is made, as it could alter the entire course for several years. This change may appear as a brilliant CEO reaching retirement age or an unsuccessful Managing Director being asked to leave before fulfilling their contract term. Regardless of the cause, a selection must occur in which a suitable successor is chosen and put into place.

All organizations strive for sustainability; they yearn to move on from one generation to the next, continually being successful. It is an understood objective for organizations to want to live forever, but very few fulfill this objective. For example, the average lifespan of a company listed in the S&P 500 index of leading US companies is 15 years (Gittleson, 2012). Sustainability depends on several factors; organizations must know how to continually evolve and adapt to meet the needs and requirements of new generations and a changing customer base. Technology is also continually changing, and organizations must be prepared to implement and utilize these new technologies whenever necessary. For example, an organization like Kodak, whose business was dependent on camera film production, has filed for bankruptcy due to the overwhelming boom of digital photography (Spector & Mattioli, 2012). Telephone landlines are becoming obsolete with the rapid expansion of cellular phones and the development of Voice-Over IP technology. Even standard walk-in retail stores quickly lose business because of the wide selection of online retailers.

But it takes more than just adapting to make a business last forever; organizations must have an excellent strategy to put them on the right path and ensure their success. The organizations' leaders develop these strategies, which

DOI: 10.1201/9781003321170-1

are only as good as the leader can envision for the business. The organization's leader carries the burden of making the toughest decisions. As a result, they are considered the face of their organization, and their impact is unparalleled. Leadership is an entity of extreme value yet lacks any true state of being. Many experts worldwide and at different universities have attempted to build on a solid explanation of true leadership. Leadership is the process of influencing others to achieve group or organizational goals. Leadership is a critical tool for success or failure for groups, organizations, and societies. If it succeeds, its citizens prosper, and when it goes wrong, teams, armies, organizations, and societies suffer (Thoroughgood et al., 2018). One of the more practical leadership definitions is that leadership is the ability or power of a leader to guide and motivate others in their work to achieve certain goals. Also, leadership can be defined as the ability of the management in an organization to create and achieve challenging objectives, react quickly and effectively when required, and inspire staff (Rajoria et al., 2022). As accurate as those definitions seem, they do not tell us much. So how does a leader efficiently guide their organization toward success?

Some leaders are considered to have single-handedly put their organization out of operation. Others are praised because they implemented a vision that saw their organization rise to the top. Finally, some organizations are identified by their leaders, those still running them, and those who have left them.

People line up by the thousands outside of Apple stores because Steve Jobs had a vision for innovative products that fit well into people's lives. Even though he was known to have ruled Apple with an iron fist (Martin, 2012), many still consider him the greatest CEO of the last 20 years, if not the greatest CEO of all time (Groth & Bhasin, 2011).

The old cliché states that not every manager is a leader, but the more important part of that sentiment is that to be a *good* manager, one *has to* be a good leader. *This is because good* managers do more than manage. They have to lead by inspiration, they have to lead by example, and they have to lead through the best times for their organizations as well as the absolute worst times.

Managers have so much to do; they carry a burden that is distributed to their subordinates, but ultimately they are the ones that are held accountable for their decisions. This is why organizations do their best to hold on to good managers. For example, great CEOs at the biggest companies in the world are given great salaries and excellent benefits, including stock options, to ensure that they will stay and not transition to a competitor with a better package. It is very hard to find a good replacement for a great leader.

Time and time, organizations have struggled to develop a good transition plan for their most important positions. A great example is the departure of

Mark Hurd in 2010, former CEO of Hewlett-Packard, who, after five years of successful management, was forced to resign when allegations that he was having an affair with a marketing consultant began to surface (Goldman, 2011). HP went through many transitions and various trials and errors before they could find a formula that worked for them. Their history represents a perfect case study for the struggles faced by organizations when attempting to manage executive transitions.

Organizations worldwide have tried to find the best process for choosing a leader. Some organizations follow their organization's hierarchal structure hoping direct subordinates of the previous manager will be knowledgeable enough to take over. Many other organizations have a risk management strategy covering the sudden departure of their top executives. They may use a temporary acting manager until they hire somebody new for the job.

No one organization has been able to find a tried-and-true universal method for transitional management or succession planning. As a result, the same organization can experience multiple results simultaneously.

EXCEPTIONAL LEADERS

Selecting a leader is also about monitoring and closely watching this leader once they take up this new role. How does an organization know it made the right decision? First, they need to examine the leaders' actions closely. What are the first things the leader does? What decisions are being made? Does the leader go through an onboarding process? Next, they will need to learn and study the corporate landscape. Where does the organization succeed, and where does it fall short? Are they new to leadership, or is this just an expansion of responsibilities? They may be required to participate in leadership training and product training. Maybe they just need to be mentored through the early days of this new role. But, once they have settled in and started taking action, this is when we need to watch very closely, and it's especially in these early days that close observation could help us understand what the potential out could be. Consider one of the most comprehensive studies conducted by McKinsey, where Birshan, Meakin, and Strovink (2107) reviewed the actions of 600 CEOs across companies listed on the S&P 500 to find out what makes an *exceptional* CEO. The results were spectacular. How did McKinsey define an exceptional CEO? They looked at the top 5% in their sample, for which their companies' returns had increased by more than 500% throughout their leadership. I think it's fair to state that if a leader delivers those kinds of results, nobody will question whether they are truly exceptional. To understand how these executives achieved these results, they closely looked at the decision-making. Still, interestingly McKinsey didn't start by looking forward and looking at the first set of actions taken – sure, they eventually got to that piece, but they looked back when they first started. Instead, they looked at the CEOs historical information. Specifically, they looked at whether they were hired internally via promotion or if they were external hires.

Interestingly, McKinsey found that exceptional CEOs were twice as likely to have been hired externally than internally. That doesn't mean that an internal hire couldn't be exceptional, there were many of those, but external hires often come in with a fresh perspective and bring a certain set of experiences that they navigated at a previous organization. External CEOs also make more aggressive decisions at the start of their career as an organization's new leaders. Setting aside poor decisions or aggressive actions to try to "prove themselves," an outsider may be able to identify gaps within an organization much faster than someone who has been part of the internal grind for a specific period. This leads to a quick decision to rectify and close up that gap. Another thought is that an external hire may come in with the idea that what they did at a previous organization worked and was successful. So, there's no reason it wouldn't work at the new company. (But, many times, there are several reasons why it could not work at the new company, but that doesn't always stop them.)

The next items McKinsey looked at were the decisions made by new CEOs within their first years at a company. Interestingly, the exceptional CEOs were 60% more likely to conduct a strategic review, allowing for a clearer perspective when it was time to make moves and take action. Frankly, it is surprising that more CEOs do not do this.

The other decision often first taken by an exceptional CEO was the implementation of a cost reduction program. And unlike the first one, the fact that this is not necessarily more common does not surprise us. Implementing a cost-reduction program is one of the hardest decisions that a leader can make. Budgets get slashed, impacting projects, availability of tools and resources, how work is sometimes conducted, and, unfortunately, a workforce reduction. But as hard as it could be to implement, a cost reduction program can have an extraordinary impact on an organization. Whether profitable or not, generating more cash flow for an organization opens the door for reprioritization, refocusing, and investing in initiatives that companies once couldn't. The hope for cost reduction programs is always to implement something that generates an impactful result but doesn't break down morale and shutter creativity to a halt, there's a fine balance for effective execution, and this is what exceptional CEOs can do so well.

Cost reduction programs worked for Mark Hurd during his tenure as the CEO of Hewlett Packard, where he turned the company into an absolute industry powerhouse following the days of Carly Fiorina. Mark Hurd began his mission at HP with an economic plan; he wanted to see the company's expenses aggressively cut down. This paid off, as in just a few years, Hurd saw HP's revenue, $118 billion in 2008, surpassing IBM's $104 billion.

Under Mark Hurd's leadership, HP's stock gained $50 billion and 6% in sales during the recession of 2008. But, those gains were all lost only one year after he was forced to resign (Koploy, 2012). But, as great as things were under Mark Hurd's control, certain departments were lagging. The Enterprise resource planning division had been overshadowed by the PC and Printer division, and if HP wanted to compete more aggressively with a company like IBM, they would have to change this. At the same time, with all the cost-cutting initiatives that Hurd was putting into place, the Research and Development division was starting to lose some strength. Despite HP's success, there wasn't complete unity amongst all the employees. With Hurd's downsizing efforts, many employees got laid off, and those still around were often stuck working with limited resources and a very tight budget. This continually hindered their creativity and freedom from making decisions.

There is a fine line that we mentioned earlier, which often has to be straddled, and this is what decides to reduce costs extremely difficult. However, as more cash becomes readily available, can an organization take advantage of this gain before employees are left functioning with minimal resources and a broken spirit?

Another example of aggressive cost-cutting initiatives is the case of Carlos Ghosn of Renault-Nissan. Carlos Ghosn spent 18 years working with Michelin and climbing the ranks until he became the head of Michelin North America. Then, he decided to switch careers and joined Renault as the Executive Vice President of Advanced Research & Development, Manufacturing, and Purchasing.

Ghosn had implemented several cost-cutting initiatives that saw Renault through some incredible turnarounds. His decisions helped drive higher profit margins throughout his career. This earned him the nickname "Le-Cost-Killer, " but more importantly, it drove the heads of the company to ask him to move to Japan and take over the role of Nissan's Chief Operating Officer.

Ghosn had four main goals he wanted to work on with Nissan. The first was the development of new automobiles and markets, the second was improving Nissan's brand image, the third was reinvesting in research and development, and the fourth was focusing on cost reduction. Unfortunately, his calculated decisions led to the closing of five factories and the reduction of 14% of the company's workforce, which did not sit well with the media at the time. Still, it was a decision that Ghosn stuck by.

As he did at Renault, Ghosn implemented tough cost-cutting programs at Nissan. His efforts in Japan were referred to as the Nissan Revival Plan, achieved in two years, one year ahead of schedule. After his work at Nissan, four plants produced automobiles based on 15 platforms, as opposed to the seven plants which produced 24 platforms. As a result, the company also saw a 20% reduction in purchasing costs (Millikin & Fu, 2005).

Compared to ordinary CEOs, exceptional ones focused first on the previously mentioned items, strategic reviews, and cost reduction programs. They focused less on this, such as mergers or acquisitions, new business product launches, and management reshuffling. Some of these were more critical in different situations, like if a CEO was coming into a low-performing organization. A wider review by McKinsey of the data goes back to the original point. The value of the outsider perspective allows fresh eyes to take a more honest look at a business. It allows for an effective strategic review to understand the landscape, make key strategic decisions, and focus on cost reduction to drive cash flow and create opportunities for reprioritization and reinvestment.

THE CHALLENGES OF LEADERSHIP

Our literature review of leadership studies from 2010 to 2017 (Elattar, 2014; Almalki, 2016; Davis, 2020) has identified that leadership as a field of study has four important challenges. These challenges are as follows:

1. How do you become a leader?
2. How do you create leaders?
3. How do you classify leaders?
4. How do you select a leader?

According to our surveys, the one that is less studied is "How do you select a leader?" As was stated earlier, the dilemma regarding leader selection is that it is not a concrete topic that can easily be represented. The areas and concerns that arise are often very ambiguous or may revolve around personal characteristics that cannot be easily measured. This leads authors and surveyors to speculate and make educated guesses about what exactly is going on when an organization reaches a point where they need to change its leadership.

The literature talks about the importance of succession planning, the dynamics between the stakeholders involved, and things such as mentoring programs and on-the-job training. Still, it gives them no tangible value. Moreover, it provides organizations with no methodologies to visualize the grand scheme of this problem and no tools to find a solution based on quantifiable data.

Much of the research may focus on one or two important factors. In contrast, organizations need to look at them together and understand how each

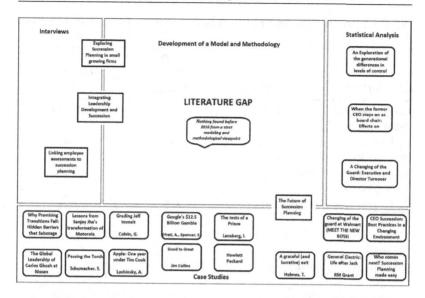

FIGURE 1.1 The literature gap

factor may positively or negatively impact the other factors. As a result, stakeholders start having the right discussions, and the organization comes closer to choosing a new leader.

We saw that today's available research uses interviews, statistical analysis, and various case studies to validate their points and reach conclusions. Still, very few develop methodologies or models to aid in visualizing the problem. Even then, these models are not entirely conclusive. They do not help us find a definite quantitative method to select the future leader.

The diagram below shows some primary research articles studied and where they fall compared to the literature gap. The literature gap in the area that this book aims to fill while utilizing some of the tools available in the other more saturated areas. In addition, this book aims to present a model which maps out the problem of leader selection, clearly understanding where the challenges lie in determining a suitable successor under certain circumstances (Figure 1.1).

NEW LITERATURE SURVEY

There have not been as many literature reviews regarding methodologies for leader selection. Therefore, our preliminary searches covered until 2016.

TABLE 1.1   Methodologies for Leader's Selection Literature Published between 2017 and 2022

AREA OF APPLICATION	DATABASE	RETURNED JOURNAL ARTICLES	RANGE OF DATES
Leader selection methodology	ABI/INFORM	118	2017–present
Leader choosing methodology	ABI/INFORM	21	2017–present
Leader picking methodology	ABI/INFORM	2	2017–present
Leader selection process	ABI/INFORM	278	2017–present
Leader choosing process	ABI/INFORM	47	2017–present
Leader picking process	ABI/INFORM	19	2017–present
Leader selection methodology	APA PsycInfo	167	2017–present
Leader choosing methodology	APA PsycInfo	6	2017–present
Leader picking methodology	APA PsycInfo	0	2017–present
Leader selection process	APA PsycInfo	167	2017–present
Leader choosing process	APA PsycInfo	24	2017–present
Leader picking process	APA PsycInfo	1	2017–present

Note: The gray shade indicates the other database used APA PsycInfo.

Therefore, a new search had to be modified to include documents published between 2017 and 2022, as shown in Table 1.1.

This new literature performed a systematic review utilizing PRISMA guidelines (Preferred Reporting Items for Systematic Reviews and Meta-Analyses) to look for leadership selection methodologies. According to Liberati et al. (2009),

> A systematic review attempts to collate all empirical evidence that fits pre-specified eligibility criteria to answer a specific research question. It uses explicit, systematic methods that are selected to minimize bias, thus

TABLE 1.2 Methodologies for Leader's Selection Searched

Leader selection methodology	Leader choosing methodology	Leader picking methodology
Leader selection process	Leader choosing process	Leader picking methodology

providing reliable findings from which conclusions can be drawn and decisions made.

Therefore, it is important to utilize the PRISMA guidelines to create and document a systematic review to ensure accuracy and good data gathering.

Besides the 850 articles reviewed from the two databases considered here, two additional papers were found using one of the papers with more citations in this area, "Are You Picking the Right Leaders?" by Sorcher and Brant from the Harvard Business Review (2002). One hundred and sixty-two articles have cited this article. Therefore, we investigated the 162 citations. From the 162 papers, only two were related to the area of leader selection. The keywords were obtained from articles such as Bennis and O'Toole (2000), Sorcher and Brant (2002), Brant et al. (2008), and Vinkenburg et al. (2014). The terms searched in the databases were as follows (Table 1.2).

These terms were searched on the databases such as ABI/INFORM and APA PsycInfo as recommended by expert librarians. During the search, a list of over 850 journal articles, conference articles, dissertations, and reports was returned (it is important to say that only the word "leader" produced a list of 141,958 documents using ABI/INFORM and 21,563 documents using APA PsycInfo).

First, the list was scrubbed to identify and remove all duplicate articles and articles that did not fit the criteria for this literature review by reading the abstracts. The list was then narrowed down to the only top two documents dealing with the leader's selection. They supported and justified our approach's uniqueness and the book's reason. These two documents can be summarized as follows:

Charles E. Naquin and Terri R. Kurtzberg (2017) have explored the selection process for leaders from the viewpoint of group cooperation. They have compared group-chosen and assigned leaders. Their empirical studies supported the idea that group-chosen leaders will be more inclined to have more of the social dimensions that assigned leaders. This social dimension will help for more group cooperation with higher trust.

Everett Spain (2020) explains that the US Army has struggled with the selection of leaders, "it has historically chosen battalion commanders, a linchpin position, based on 90-second file reviews." Therefore, in 2019, the US Army revamped that selection process, which now includes a thorough physical, cognitive, and psychological assessment and interview assessment. This new process is good for hierarchical organizations that have been members of that organization but not for external candidates (in particular, in the business world).

THE DIFFERENT CHAPTERS OF THE BOOK

This remarkable book discusses an important topic in leadership: How to Select your leader. The authors identify the application gap and present a methodology based on Total Quality Management (TQM) to support the guidance of a process to select leadership (at any level of the organization).

The book is organized into five chapters. A synopsis of the chapters follows:

- Chapter 1 – Investigating the Leader Selection Process: This chapter introduces the topic of leader selection and our efforts to discover prior work. This challenge in leadership is less studied. Finally, the chapter concludes with a recent survey of the last five years (2017–2022) that justifies the uniqueness of this book.
- Chapter 2 – Classification of Leadership Styles: There are many classification schemes for leadership. The authors introduced and justified the framework adopted for this book, which has four different categories: cultivating, inclusive, directing, and visionary leadership styles.
- Chapter 3 – The Matrix of Change and Total Quality Management: The Matrix of Change (MOC) derived from Total Quality Management (TQM) is introduced in this chapter. TQM is a philosophy with many tools to guide the design of high-quality new systems. Those are the basis for the MOC (developed by the Massachusetts Institute of Technology). We consider that the MOC can be the backbone to guide the selection process to select a leader.
- Chapter 4: Case Study: Yahoo!: The utilization of the MOC for leader selection is explained with a case study in Chapter 4. A case study of the 2010s helps us visualize how the MOC deals with complexity. The case study of Yahoo in the early 2010s is utilized for this.

- Chapter 5: Advances in Leadership Development: This chapter introduces the next frontier in leadership: Virtual Reality/Virtual Simulation (which has become very due to the Metaverse). Under the premise that experiential learning is essential to building leaders and the closest thing (beyond textbooks and case studies discussions) to experiential learning is Virtual Reality/Virtual Simulation, the chapter goes through experimental analysis to show some initial results.

Classification of Leadership Styles

2

Numerous classification systems have been utilized to analyze, define, and measure leadership styles and identify personality and leadership attributes. Tests connected to these classification schemes can provide a wealth of information about the existing leadership styles. They can also be used to quantify leadership development progress. We have studied several classification methods, but the one that fits our work is the one from Charles Pellerin (2009).

4-D LEADERSHIP SYSTEM

The 4-D leadership paradigm is a valuable asset in today's leadership literature. Charles Pellerin (2009), a former director of the Astrophysics Division of NASA, created this leadership classification. His ability to think creatively about leadership originates from his involvement in the Hubble space telescope disaster. There's nothing unusual about having a bad day at the office. But some people have worse days than others; at this time, Charles Pellerin had a few notable ones. Not many people have to explain why an organization has invested a decade and a half and $3 billion on a failed project.

The biggest issue with the Hubble telescope, launched in 1990, was a faulty mirror that had to be fixed in orbit, which had what appeared to be an unfixable flaw in its optical system. The cause of the problem was not so much a technical fault with the contractor but a lack of leadership. Unfortunately, NASA's connection with the contractor was so abrasive that it prevented the contractor from reporting this technical issue. Even though he successfully repaired the Hubble, he was deeply troubled by his association with the original leadership crisis. As a result, he made it a point to focus his book on how

DOI: 10.1201/9781003321170-2

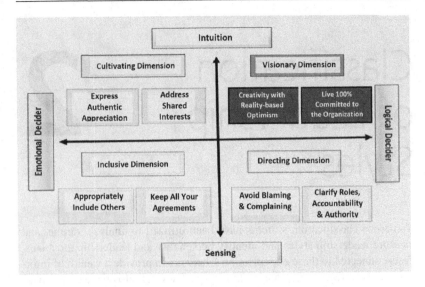

FIGURE 2.1   The 4-D System with the four dimensions of *Cultivating*, *Inclusive*, *Visionary*, and *Directing*

to address the problem of many CEOs forgetting social context when managing initiatives.

Pellerin's (2009) 4-D System was built on the social component of leadership. "Social aspects influence human behaviors, and so influence a technical team's potential to perform or not," Pellerin stated as his main belief. A 4-D leadership approach can also guide "to lessen or eradicate social context risk from your team." He went on to say that the factor of social involvement affects behavior and perception.

As indicated in Figure 2.1, Pellerin identified four types of leadership styles:

- The *Visionary* leadership style has logical and intuiting dimensions. This style relies on the thinking of future opportunities and is considered influential/charismatic leaders who often create what they want.
- The *Cultivating* leadership style relies on emotional and intuiting dimensions. This style encourages empowering feelings, caring about others sincerely, and achieving a better world to live in.
- The *Inclusive* leadership style depends on emotional and sensing capabilities. This is someone who communicates and places emphasis on building relationships with people.

- Finally, the *Directing* leadership style drives decisions by utilizing logical and sensing dimensions. This style encourages taking action and directing others, including managing plans, organizing, and taking control.

The 4-D Leadership model comprises two basic dimensions: the y-axis, innate information preference (sensed information and intuitive information), and the x-axis, inherent choosing preference, as shown in Figure 2.1 (emotional decider and logical decider). The 4-D leadership approach was validated in Kouzes and Posner's study of leadership effectiveness which included the following (Kouzes & Posner, 2007):

a. a 1,500-person survey by the American Management Association,
b. a follow-up study of 80 senior executives in the federal government, and
c. a survey of 2,600 top-level executives who completed a checklist of common leadership traits. Only one question was asked: "What do you admire the most in people?":

- Eighty percent of those polled favored being a trustworthy leader. This response emphasizes the need to be open and honest with others while still ensuring inclusivity. This description closely resembles the *Inclusive* style.
- According to 67% of those polled, leaders should be productive and efficient. This leadership style is the most compatible with the *Directing* style.
- The most significant leadership attribute, according to 62% of those polled, is forward-thinking leadership. This description closely resembles *Visionary*.
- An important attribute is a concern for the well-begin of others; according to 58% of those polled, it is special. Leaders who follow this leadership style are frequently concerned about the well-being of others. This is quite similar to the *Cultivating* leadership style.

  Each quadrant is also broken down into two behaviors (eight total). All these actions relate to the five practices identified by Kouzes and Posner. These eight actions include:

  – Genuinely express gratitude and mention common interest – According to Ankush Joshi, service line manager at Informix USA, "mailing a human thank-you note, rather than sending an email, will do wonders." Building a successful team can be greatly aided by demonstrating respect for your team members. People from various backgrounds can be found across an organization. Therefore, it will take time for them to all come to share the same values and ideas. The cohesiveness among a team could be increased through team-building activities. Aligning their values and interests would be the first step. Open communication between lower management and the workforce can continue this process until shared values materialize. Their shared views and common ideals are the strongest ties between leaders and followers. Leaders must connect their values with those of the people they seek to influence to obtain their organization's complete trust.

– Consider others appropriately and uphold all your commitments. Other people's inclusion is like Kouze's attitude of "allowing others to act." Kouzes and Posner assert that effective leaders must promote teamwork and empower others. Leaders must foster a culture of trust, be open to influence, and enable interactions to foster collaboration. An essential part of leadership is fostering a culture of trust. When you have demonstrated that you will keep your word and honor commitments, are sincere in your interactions with others, and are not concerned that they will suffer punishment for giving you unfavorable criticism. Then, an employee will trust their employer.

– Be devoted and show reality-based optimism. Leaders need to be visionaries who are constantly willing to change their viewpoints. They must also be able to inspire others to achieve their objectives. Leaders must set an example and exhibit the values they want to see in their followers to demonstrate that they are fully committed. Leaders must "personify the common principles" and educate others to replicate the values, "according to Kouzes' Setting the Example." Regardless of your position, be it the president of a nation, CEO of a business, or even shift leaders at work, followers will always look to you to set the example and model the behavior you want from them.

– Of the eight behaviors, the final two are the most at odds with the findings of Kouzes and Posner's research. Avoid blaming and whining, and clear responsibilities, accountability, and authority. Kouzes makes it abundantly obvious that you cannot, however, "allow others to act" without also holding them responsible.

The 4-D process is relatively like existing protocols (such as the Myers-Briggs) used for personality/temperament and individual/group compatibility training. For example, millions of people have taken the Myers-Briggs Type Indicator (MBTI), developed by Katherine Briggs and Isabel Briggs Myers and released in 1962 (Briggs-Myers et al., 1998). It is based on Carl S. Jung's archetypes.

The MBTI instruments identify preferences in four important domains:

1. Where a person focuses their attention
   - Extraversion (E)
   - Introversion (I)
2. The way a person gathers information
   - Sensing (S)
   - Intuition (N)
3. The way a person makes decisions
   - Thinking (T)
   - Feeling (F)
4. How a person deals with the outer world
   - Judging (J)
   - Perceiving (P)

The 4-D test has been applied in public and private organizations, with positive results for managers, supervisors, and employees. However, compared to the Myers-Briggs classification method, the 4-D leadership system provides a precise, accurate, and approachable method of evaluating leadership. The 4-D coordinate system, according to Pellerin (2009), "rearranges the core components of high-performance organizations and compelling leaders." Jung's personality types of taxonomy inspired the 4-D system and the Myers-Briggs Test Indicator (MBTI). The 4-D leadership method divides leadership into four categories that are easily assessed. The MBTI, on the other hand, uses a 16-style system to describe leadership and personality, making it difficult to track improvements (See Figure 2.2).

Another key factor to consider when choosing an evaluation system is whether it can effectively assess leadership for teams and projects. Any project team can benefit from 4-D's ability to identify the leader. For example, Raytheon Corporation's previous vice president and general manager, Dr. Anthony Calio, stated:

- The 4-D evaluation technique, refined over 15 years of working with NASA venture, engineering, and management teams, has the potential to improve team execution in almost any project. In addition, it emphasizes the importance of comprehending social

ISTJ	ISFJ	INFJ	INTJ
Responsible Executors	Dedicated Stewards	Insightful Motivators	Visionary Strategists
ISTP	ISFP	INFP	INTP
Nimble Pragmatics	Practical Custodians	Inspired Crusaders	Expansive Analyzers
ESTP	ESFP	ENFP	ENTP
Dynamic Mavericks	Enthusiastic Improvisors	Impassioned Catalysts	Innovative Explorers
ESTJ	ESFJ	ENFJ	ENTJ
Efficient Drivers	Committed Builders	Engaging Mobilizers	Strategic Directors

FIGURE 2.2   The 16 personality types (Myers-Briggs psychological test)

standards and behaviors, which are sometimes overlooked in specialized businesses yet can be critical to their success.
- Pellerin further stated that the 4-D technique has been utilized in over 2,000 workshops and that its accuracy is over 90%.

CASE STUDY OF GE AND ITS TWO CEOS: WELCH (1981–2001) AND IMMELT (2001–2017)

All four dimensions can be found within an organization, but usually, one stands out more than the others and influences how the company is perceived. The culture that is most prevalent in the 21st century is the *Directing* culture. For example, in General Electric (GE), since Immelt became GE's CEO, a

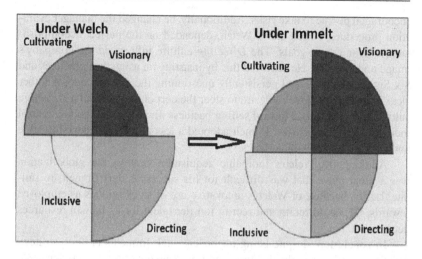

FIGURE 2.3 Comparisons between Welch and Immelt

more balanced leadership became more prevalent. Figure 2.3 illustrates this departure from the Welch era, as discussed by our research group (Laval et al., 2012).

The *Visionary* culture experienced the most growth during Immelt. Under Welch, GE used inorganic growth strategies, which involved numerous acquisitions. However, Welch also emphasized the significance and benefits of elastic short-term goals. The CEO of GE, Jeff Immelt, tasked his corporate executives with finding new "growth platforms" that could provide high-level operating profits in a few years. Immelt adopted a plan that relied on commercial excellence, technological superiority, and worldwide growth.

Under Immelt, the *Inclusive* culture has also experienced a significant increase. While Welch adopted several *Inclusive* tactics, such as reducing the organization chart, implementing 360-degree feedback for managers and executives, and using trust sessions, Immelt has made many more adjustments to expand this aspect of the business culture (Laval et al., 2012). Immelt underwent the most significant transformation when the company's perspective on marketing changed. Immelt told his senior managers to use marketing as a powerful instrument.

Under Immelt, the focus on *Cultivating* and *Directing* cultures decreased the most. Under Welch, the *Cultivating* dimension was fueled by regular job rotations to give staff a greater understanding of the company. Welch used training facilities and numerous training plans to develop his team. Immelt enhanced GE's attempts to recruit from outside firms, decided to slow job rotations to develop leaders with more subject experience, and continually encouraged people to take risks. Additionally, he changed the manager evaluation procedure, which under Welch, depended on the perfect accomplishment of immediate goals. The *Directing* culture influenced the company's image under Welch. He achieved this by insisting on using best practices and Six Sigma standards and persistently questioning the existing quo. Another clear indication of Welch's desire to steer the corporation toward a *Directing* culture was his attitude toward selling business divisions that did not control their respective industries. Immelt adopted a less *Directing* leadership style but has kept some characteristics.

Additionally, Welch's inorganic acquisition strategy for globalization was a great move that was difficult for his successor, Jeff Immelt, to imitate. Lastly, because of Welch's innovative use of stock options as employee rewards, GE could retain and recruit top-tier talent for its human resources department.

In conclusion, despite being unconventional, Immelt's organic strategy was reasonable, given the dynamic and fragmented nature of unproven start-ups with inflated prices (Laval et al., 2012). However, he also inherited a two-dimensional organization (*Directing* and *Cultivating*) devoid of the other two dimensions (*Visionary* and *Inclusive*) needed for adjusting to and succeeding in the changing external environment (Laval et al., 2012).

Immelt was enthusiastic about developing initiatives, particularly for the *Visionary* dimension (investing in research and development). He even came up with the phrase "Imagination Breakthroughs" (IBS), which were described as innovations that "put growth on steroids," were "game-changers," and necessitated a "big swing" (Bartlett et al., 2007). In addition, Immelt occasionally used storylines to refute the idea that GE's growth peaked at the turn of the century, saying, "in the late 1990s, we became business traders, not company growers" (Bartlett et al., 2007). Finally, Immelt employed storylines, saying, "Before we launched our growth initiative, marketing was a place where washed-up salespeople went" (Stewart, 2006).

Therefore, the 4-D system can be used to study transitions and even to support the selection process for a new leader based on the previous successes/failures of the organization. In addition, it can help to see who can fit more with the trends of technologies, the economy, globalization of the markets, and what needs to be changed.

The Matrix of Change and Total Quality Management

3

Choosing a leader often comes with change. Is there an outgoing leader? Is this the new leader's first opportunity in such a role? And what is the perception of those individuals who will be working with this new leader? Although the hardest thing about change is often uncertainty, these questions are only a minor glimpse of the uncertainty built up when a new leader is about to take the helm. Uncertainty holds back decision-making, sometimes putting it at a complete standstill. One is not if they can maneuver the consequences, and when it comes to business, one is often not sure if they can afford the consequences.

Unfortunately, there is no way to predict outcomes and different consequences of certain decisions, like selecting your leader, but experts often work toward ways to minimize negative results. We learn from experience and past actions. This process is a way to shape our decision-making. Historical references and data points are often used to help guide the way when navigating unpredictable circumstances.

One can always take the gamble, throwing a *Hail Mary*. Those actions stick sometimes, but it's often a result of luck than any scientific calculations. And when the stakes are high, taking the gamble will not cut it more often than not.

These situations are where modeling and simulation come into play. Remember, we have stated that trying to predict the future is impossible, so the next best thing is doing our very best to minimize negative outcomes due to our decision. Thousands of techniques can be used to model and simulate the outcomes of our decision-making process. But the ones we believe truly have a beneficial impact are those that allow stakeholders to look at the big picture and generate discussions.

DOI: 10.1201/9781003321170-3

THE MATRIX OF CHANGE (MOC) (ELATTAR, 2014)

Sometimes when individuals are deep in the decision-making process, they develop blinders, preventing them from understanding all the impact of their decision. They may have a sense of cause and effect. Still, when an organization is big, there are many different organizations, product lines, suppliers, partners, etc. That may impact this decision, which is why open dialog needs to be promoted. In addition to seeing the big picture, organizations must understand what is needed to commit to change. It is an investment to change. It calls for resources, tools, headcount, and a mission everyone can buy into. Organizations have to find a way to model the level of difficulty such a change brings. It is not just an analysis of what needs to change, nor how fast it needs to change, but what is required to perform this change and make it happen most efficiently.

Think of an athlete coming back from an injury. Say a football player who broke his leg. They may be sitting on a chair with their leg in a cast at that very moment. That is their current state. In their mind, the future state is to be back on the field, running for a touchdown pass at a championship game. But what needs to occur in that space between them, sitting with a broken leg wrapped up in a cast and running for a touchdown pass? This transitional space in the middle is where conditioning lives, where rehabilitation, physical therapy, and re-strengthening of the muscle all take place. The athlete must understand all the work and investment needed in that transitional phase to return to the field again.

This transitional space is the area we seek to model because it is an area that is often overlooked. Experts write about the changes needed for success, point out faults in the existing process, and discuss the gaps today. We also see certain leaders' or stakeholders' downfalls and why they fall short. Experts also write about the future state, where things should be, and how things should be, and they discuss why an organization could be doing so much better if it was within that future state. And oddly enough, despite us not being able to predict the future, we seem to think that things could be so much more efficient if we were within a different stage. But the area often overlooked is the space in the middle, the change, the investment, the requirement, the difficulty level, or the ease level. Here is where I am today and where I want to be in the future, but we need to talk about what it will take to get us there, and one tool that helps with such a process is the MOC.

The MOC was a joint research project developed by the Massachusetts Institute of Technology's Center for Coordination Science and the Center for eBusiness@MIT (http://ccs.mit.edu/moc/). Intel Corporation and British Telecom predominantly funded it. Dr. Erik Brynjolfsson was the genius behind the idea (Brynjolfsson et al., 1997). The MOC is a method developed to model change management. It identifies complementary and interfering work practices and comprises three interconnected matrices: the existing practices, the target practices, and a transitional matrix connecting the first two (see Figure 3.1).

The first matrix is that of an organization's existing practices. It allows stakeholders to visualize the practices that comprise where an organization currently stands and gives them weight to rank their importance. It provides the stakeholders with an understanding of whether these practices positively or negatively impact the organization. It will also examine the relationship between these practices and determine whether they interfere with or complement one another (Massachusetts Institute of Technology – http://ccs.mit.edu/moc/).

The second matrix represents an organization's target practices and represents where an organization wants to be in the future. It may or may not

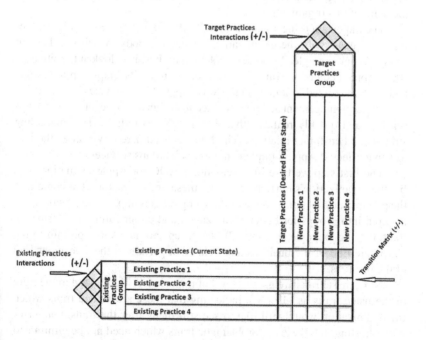

FIGURE 3.1   The matrix of change

contain practices already displayed in the existing practices and, more often than not, will display newly introduced practices, depending on the organization's objectives. Like the existing practices, the target practices will also be given a ranking and compared with one another to identify their relationship dynamics.

Finally, the third matrix is a transitional area, identifying the relationships between the existing and target practices and showing whether they interfere with or complement each other. Generally, if there is a large amount of interference between the existing and target practices, this usually indicates that the transition may be difficult. But, in contrast, if the majority of practices complement each other, this is usually indicative of an easier transition.

One advantage of using the MOC is that it starts a dialog and allows for stakeholder discussion. Once they see what practices are important to them and how they impact the organization, they'll also start to appreciate those practices that do not pertain to them and may help them make more sound decisions.

The MOC is a method that has been applied in several different fields, such as healthcare, manufacturing, and retail. But in our case, we apply the MOC to the organization's dynamics and the leader we would like to see at the helm of this organization.

The important thing about the MOC is that we can better visualize aligning those goals with the organization's managing body, which will have to change in our example. A good example of the MOC is to look at the situation when Apple decided to transition from outsourcing its Maps app to Google and instead decided to implement its own Apple Maps services.

Senior management at Apple decided to design a new generation of Maps services and quickly realized that they needed to rethink their marketing strategy and customer relations. The MOC is used here to visualize the factors that affected Apple's implementation of its Maps services.

The first step we take in developing a MOC for Apple's situation is to list the organization's current practices; these are the factors that come into the picture due to Apple's usage of the Google Maps app. These practices are also put into groups for better organization and easier future discussion; in the following example (Figure 3.2), the groups are items that pertain to the Corporate End of the business and items that pertain to the Consumer End of the business.

Next, these traits are given a level of importance, highlighted to the right of the listed traits as +1 or +2, indicating that these are valued traits which the organization would still like to see still there after the transition, −1 or −2, indicating that these are problematic traits which need not be eliminated when the transition takes place, or 0, indicating that there is no real preference regarding this trait during the transition (Figure 3.3). Then, on the right side

Utilization of Google Maps	
Consumer End	Customer's utilizes familiar map service
	Synchronization with Google Maps desktop version
	Stronger overall integration of third party software
	Weak Integration with iPhone, iPad
Corporate End	Google gains customer usage information
	iPhone user info used by Google to improve Android OS
	Apple pays Google licensing fees
	Apple unable to customize Google Maps

FIGURE 3.2 Organizing the existing traits/practices

of the listed traits, we begin to compare these traits to one another, observing whether they complement each other, indicated by a "+" sign, interfere with each other, indicated by a "–" sign, or whether the organization is unsure how a certain trait impacts another one, in which a "?" is assigned.

The same procedure is then applied to develop the future state. Finally, all the desired traits are listed on the upper right side of the matrix and organized into general groups. This is visualized in Figure 3.4.

As was done for the list of current traits, the desired traits will be ranked, and their importance level will be noted. We will also begin to compare all the desired traits to one another to determine whether they interfere with or complement each other. Their status will be highlighted by either a positive or negative sign as displayed in Figure 3.5.

At this point, we can see the full list of the current traits resulting from Apple's utilization of Google Maps and how they interact with one another.

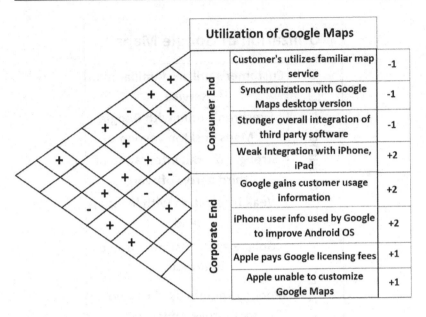

Utilization of Google Maps		
Customer's utilizes familiar map service	Consumer End	-1
Synchronization with Google Maps desktop version		-1
Stronger overall integration of third party software		-1
Weak Integration with iPhone, iPad		+2
Google gains customer usage information	Corporate End	+2
iPhone user info used by Google to improve Android OS		+2
Apple pays Google licensing fees		+1
Apple unable to customize Google Maps		+1

FIGURE 3.3   Including the current trait interactions and rankings

Utilization of Google Maps (Target)	Consumer End				Corporate End				
	Customers less confident using new Apple maps service	No desktop version for easy synchronization	Weak Integration with third party software	Strong Integration with iPhone, iPad	Apple maintains customer usage information	Apple continues to strengthen its own devices	Apple saves cost by avoiding Google licensing fees	Apple has full customization of own maps service	

FIGURE 3.4   Organizing the future state/practices

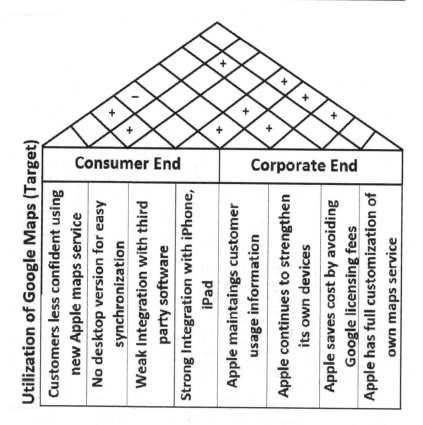

FIGURE 3.5 Including the future trait interactions and rankings

We also see the desired traits of how things will be once Apple Maps is installed as the default maps app at the end of the designated transition time. What is now missing is how the current state affects the future state. This is what is represented in the final diagram below (Figure 3.6) and what is now considered a complete MOC. Each trait from the current state is compared to every trait in the future state and studied to understand whether they interfere with or reinforce the future state. As was done with the traits in their respective states, a positive or negative sign represents the interaction between the traits.

The completed MOC displays a list of the organization's current practices and all the traits Apple desires in the future as it changes the default apps that come as part of its mobile software. These traits are compared to one another in their respective states and then compared once again across states. Seeing the MOC in its completed form, an organization can utilize it

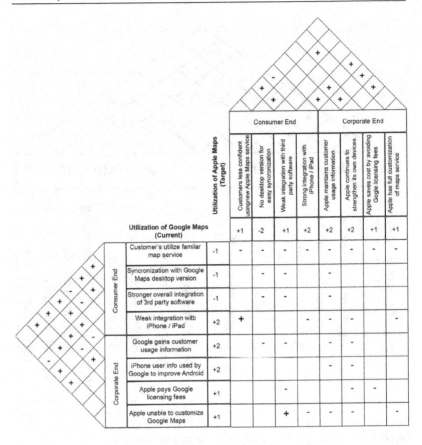

FIGURE 3.6 Adding the center matrix

as a tool to determine the scope of this transition project. If the center Matrix, where the present and future states are compared, is filled with many negative signs, the transition procedure will be complicated. When the future state differs greatly from how things are being done today, an organization will be faced with a very big learning curve that it will have to overcome. This will take a greater deal of planning, including many more compromises, and will be a greater undertaking requiring all affected parties to be on board and dedicated to reaching the desired goal.

On the opposite end of the spectrum, if the center Matrix appears to be flooded with many positive signs, this indicates that the future state is not very different from the current state. The organization is not very far from where it wants to be regarding the type of manager they have leading them. As a result, there will be fewer risks, and the transition will be manageable.

It is important to remember that the MOC does not provide a solution to problems in transitional management. Rather, it paints a picture of what the transition process may be like and allows stakeholders to understand better the undertaking required for a successful transition. Therefore, the MOC can be considered the first step of the planning process for change management.

To offer the best guidance when it comes time to change, an organization must begin with the MOC. They must list out their current organizational practices and rank and compare them to understand whether the aspects of their business, as it stands, complement or interfere with one another. Once this step has been completed, the same action must be taken for the future state. This is where companies lay out their vision of where they want to be after a certain period. Once the center matrix has been developed, bridging the two states, stakeholders are brought in to discuss the transition state. Will it be a hard change, or is the state of affairs not as gloomy as they had imagined? The value of the MOC is that it presents an overall picture. Division leaders may not be aware of the importance of certain aspects in other divisions, and it may take the MOC to bring their attention to it.

TOTAL QUALITY MANAGEMENT (TQM)

One thought to always remember is that selecting a leader is not, and will never be an exact science. No matter how many methods, models, or tools one applies, to the process, no matter how many PhDs are brought together to lay the path forward. There is no exact measurement and no immediate sign of success. Selecting a leader is more of an art than an exact science. But it is an art that can be greatly improved when the right systematic process, methods, models, or tools are applied. This goal is what this book aims to accomplish and communicate. A painter who begins a new project with a blank canvas will have an arsenal of paintbrushes of different sizes, sponges, and colors as far as the eye can see. A carpenter will show up to his job with a toolbox, sometimes two, that give him a selection of options to utilize for the task at hand. It is no different when a leader needs to be selected. This book serves as a toolbox with several options that can be leveraged together or individually to provide a process that improves and offers guidance at every step.

Another concept we'd like to introduce here into your toolbox is Total Quality Management or TQM. A framework built around delivering customer satisfaction and using a customer-centric approach drives every aspect of your business toward success.

And this makes sense. This concept is not a revolutionary one. Still, the pieces coming together to form the framework are, and it structures the components together to make it applicable. We say it makes sense because why else are any organizations in business today if not to deliver customer satisfaction? Many organizations get so weighed down with internal processes and a complete block of the peripheral visions that they lose sight of what their customers want and how valuable their input is. When organizations begin formulating their product roadmaps without considering what their customers want, they operate in a bubble. They may think they are doing what is best for their customers. Still, that belief could be an entirely outdated one or one that has shifted and changed so much from the original customer requirements that it is no longer relevant. But an organization may not realize this, and that is where the real danger can lie. When major investments are made and a product hits the market, all eyes will be on how customers react and how customer purchase. If this does not happen (at least to the stakeholder's expectations), one of the first questions will be asked: "Did you conduct any customer research?"

An organization shifting its mentality to be customer-centric is not always an easy one, despite how easy the concept is to grasp. Accepting that this model needs to be adopted is not one that comes naturally. This situation is where the TQM framework can be invaluable. Consider the eight key components of TQM:

1. Customer comes first
2. Employee engagement
3. Process at the core
4. System Integration
5. Continuous improvement   6.   Effective communication
7. Data-driven decision-making
8. Strategic and Systematic approaches for problem-solving

All of them are in place for a specific purpose and serve their function, they may not necessarily need to follow the above-outlined order, but when all is said and done, they are all there to drive the first element. The customer comes first.

Putting the customer first and setting a path to deliver their requirements effectively is half the battle. Unfortunately, we have seen leaders in organizations state that if a customer's feedback does not currently align with our existing road map, it will not be considered. And this is exactly how a leader starts to lose their way. Why try to capture the customer's feedback if you are only going to apply it if it fits your plan? This attitude is not agile, this is not effective leadership, and this is not how you put the customer first.

TQM draws its concepts from several different teachings. Its origins date back to the 1950s or even before. Still, it was in the 1980s that the framework began to be applied and was widely used in the United States Navy and several divisions of the department of defense. The concept eventually made its way into manufacturing and the corporate world and aligned with several of the current ISO standards.

The second concept of TQM is employee engagement. This concept brings together the entire staff, all workers, leadership, etc., to align on a common vision that feeds the customer-centric objective. True employee engagement opens up the door for effective collaboration and innovation. Employees need to express their ideas, how tasks can be done faster, how gaps can be identified, and how the organization's work comes closer to delivering the customer requirements. This feature is also one of the most effective ways to feed another concept within TQM, the idea of continuous improvement.

Continuous improvement is how an organization stays ahead of its competition. As they look toward their processes and strategic approaches (the other concepts in TQM we'll describe in a moment), a leader and their employees must continually observe how things take place within their facilities and their cycles. While several leaders decide good enough may be good enough (and at times it is), something can be improved more often than not. True leaders do not hide from where flaws in their systems may lie, they don't dismiss ideas that come from their subordinates, and they need to embrace the idea that those who live these processes, day in and day out, could have some of the best ideas of how situations could be improved. When encouraged, continuous improvements need to be embraced and rewarded, and employee engagement will bring this change forward.

The following process, utilizing an integrated system, and looking at a strategic approach, are key elements when it comes to identifying what the organization does right. And when the right way is discovered, putting it in place and into process offers repetition and a way to duplicate this success. In addition, the process can be an excellent way to be agile and move things quickly and effectively through the necessary cycles. But a closer look must be taken when a process gets in the way of results. An analysis of where gaps are being identified, where bottlenecks could be coming up, and from there, small changes to the process can be adjusted, sometimes that's all it takes.

How the process impacts the overall system is a good indication of whether it is working. As mentioned, an organization thrives when it can take advantage of an integrated system. Like everything in this world, the more there is collaboration, the more transparency, and the more integrated actions are, the better the outcome. But one needs to watch out for the dependencies that arise from these systems. A good leader knows when to implement safeguards and ensure the entire system does not break if a single process fails. But when the sales organization knows what is happening in the production division, the production group knows what is happening in finance, and finance knows what's happening with customer support. The systems will have fewer chances of being compromised. A good leader knows to bring these organizations together and ensure every action delivers a positive impact across its integrations.

But an integrated system will eventually fail if there is no effective communication. So the next concept of TQM is extremely critical for success: whether the organization is going through a merger or whether it is on a daily stand up to review the upcoming tasks. Effective communication is an element that needs to be embraced by everybody, from the leader at the highest point of the organization chart down to the newly hired employee finding their way across the filing cabinets. Effective communication instills trust in employees, instills confidence in the leader, and reduces confusion, complexity, and uncertainty. And if a leader does not know the right answer, that is ok. That should be communicated as well. And if this is the case, a good leader should know that this is an opportunity to bring together the employees for brainstorming sessions, process improvement exercises, or an agreement for a new path forward. This concept brings us to the final concept of TQM: the idea around data-driven decision-making.

Data-driven decision-making is how a company navigates uncertainty. An investor does not spend $10,000 on the stock of a company they know nothing about. A successful investor will do their research. They will review historical data, study earning reports, and understand the market the company has been operating it. A successful investor will let the data drive their decision, not a hunch, not a feeling, not because the company has a cool-sounding name. The data guides their path. And a leader looking to embrace TQM will do the same as they build up their organization. But as stated at the beginning of this section. Selecting a leader will never be an exact science, and making data-driven decisions is the same thing. But it brings one closer to the right decision, reduces risk, and increases the probability of success. Making data-driven decisions is how an organization justifies its actions to its shareholders. It is how an organization can identify that a process may be improved before actually implementing it. It's how an effective conversation between teams thrives and moves forward. Data-driven decisions differ from throwing a *Hail Mary* and moving the offense away from defensive tactics. A data-driven decision is what helps understand what customers truly want. With each data-driven decision, we can run a customer-centric business, meet their needs and keep an open line of communication.

A good leader takes all these concepts and puts them together into their toolbox. TQM is not the be-all, end-all that will solve the company's most critical problems. None of the lessons in this book will serve this purpose on their own. Still, when combined, when utilized at the right time and in the right situations, effective change will start to take place, and an organization will continue to improve and thrive.

Selecting a leader who does not shy away from TQM is certainly a step in the right direction. Sometimes this is a fundamental change that an organization so desperately needs. Other times it is a framework that is needed to align with how the organization currently functions and is a method to keep this from falling apart. But this is one framework in our systematic model that brings us closer to an effective selection.

COMBINING THE MOC AND TQM

TQM integrates quite efficiently with the MOC. As one thinks about effective leadership, sustainability is a concept that must always be at the forefront. Sustainability in management ensures the continuation of effective leadership. This continuation is or should be there, regardless of the selected leader. While often botched in practice, the transition of leadership is best when it is seamless. Services should not be interrupted, customer satisfaction should not decline, employee morale should not drop, and revenue should see a climb. Of course, this is a very hypothetical idea, and the concept is easier to write about than to execute. But this is why we're on the topic, perfect execution is a thing of dreams, but organizations can get very close. Think of best practices and consider all angles; where are the pitfalls? Where are the bottlenecks? Where have similar organizations fallen short, and what determines a certain company's success? These are all things that could be studied through modeling and simulation. Therefore, integrating TQM with the MOC could lead to a meaningful outcome.

The idea of integrating the two concepts is to take the eight main components of TQM and apply them to the MOC, essentially running them through the transition exercise to understand the level of difficulty when it comes to applying a change to each one of the components. This integration is important because it shows us how an organization's overall transition will be and gives us a better understanding of where it should focus its efforts and investments.

Consider the most important item of the TQM components: the customer comes first. If it's already bad, a leader's primary goal, before anything else, should be to change this. Nothing should be worked on until a plan is implemented to improve customer satisfaction and ensure the service meets market demands. If you don't have customers, you don't have a business. It is that simple. But even this single component can be broken down into several pieces, which can be reflected in the MOC. How is the customer not coming first? Are the products not functioning as they should be? Are there shipping delays when orders are placed? Is product management not effectively captured capturing the customer requirements? Is after-sales support unable to address bugs, defects, or surprises?

Each area here can be closely looked at and modeled in the MOC. If only one component of TQM should be selected for the transition exercise, this should be the one. How an organization addresses its customers' most important demands is what skyrockets its sales and lays the foundation for long-term success. This aspect grows more important today with so many social media platforms so easily accessible, reviews available in everyone's palm of their hands, and word of mouth spreading like wildfire. To use an extreme example, consider how Snapchat lost $1.3 Billion in a matter of minutes following a tweet by celebrity Kylie Jenner who was displeased with the app's new interface (Yurieff, 2018). Granted, most people don't have millions of followers to warrant Billion-dollar losses, but how many times have you been turned off by a business based on its yelp reviews or made a decision on Amazon purchases after scrolling through the listed reviews? Information accessibility has made putting the customer first a crucial task. And, in general, it's just the right thing to do.

Employee engagement is the next major component of TQM and is the next theme that can be modeled across the MOC. One of the toughest tasks of any new leader is breaking down silos internally. This situation is especially true within large organizations with many activities, several product lines, and continuously expanding departments. With rising pressure, employees get sucked into their roles and focus on delivering increases – which is often a good thing, but it may come at the cost of performing your job within a bubble. For example, engineering may not necessarily be in tune with Product Management, which is not in tune with Sales or Customer Support. It becomes a frustrating experience for a customer – and if you have not figured it out by now – it impacts the previous component of TQM. This experience tells us that employee engagement is another excellent component that benefits when it's modeled through the MOC. Still, instead of just asking what we can do to engage our employees, we think a more important question is *which* group of employees within our organization should engage with one another. It's not just about getting people to work together. It's about getting the right people working together. Would certain departments be more efficient if there were daily stand-ups with another department? How about a complete merging of two groups of the business? And then, do organizations have to follow the typical mold of having separate departments for each job function? How does transitioning such a change impact the business? Unfortunately, not enough leaders take a step back to look at the departments and ask, "Does this make sense?" – that is because it's not such an easy question to answer, and it takes a keen eye to make the right judgment call. Then there is the idea of process at the core. This one must be approached with extreme caution. The process is needed and effective, keeping the business from spiraling out of control. But too much process can hinder progress, too much process, especially process for the sake of process, can put opportunities at a standstill with missed deadlines, and an endless revolving door preventing effective decision-making.

But even with the best processes, ongoing revisiting and continuous improvements must always be ongoing. Nothing is ever perfect, and any organization that thinks they've achieved perfection must closely look at how its performance is progressing. As companies grow and businesses expand, processes become a necessary evil. They help keep things organized, bring order to chaotic situations, and keep teams from feeling overwhelmed. Still, we must ensure the process does not bring analysis paralysis. It is a fine line that organizations need to walk. How do we implement the best process? How do we improve this process in time? This case is a great fit for the MOC. Laying out the process in the MOC and running it through the exercise can help generate a conversation around bottlenecks, cost cutting, and reducing overhead. But what if there is no process altogether, this is an even more difficult problem. Within the MOC, one can outline the options to be considered and look at how difficult they would be to implement, the headcount required for such a process, and if any new tools would be needed. And then, ultimately, what are we trying to achieve with this process? Is it meant to reduce money? Time? Is it meant to address requests from customers or internals systematically? Does it improve customer satisfaction? Does it improve employee morale? Why should the first question be asked, and can the exercise occur from there?

System integration is another tricky one and maybe harder to model, as it often depends on factors out of our control. Remember, a system can be anything, but it can combine software tools, hardware platforms, and certain processes for many organizations. One of the trickiest things about integrating two or more systems is ensuring you do not break an existing system. This integration can lead to downtime, loss in revenue, and even more cost trying to fix the problem. In an ideal world, systems would integrate with no downtime, and yes, while it is possible, it takes very careful management and the right individuals involved, and sometimes, it's just a matter of luck. We have seen situations where companies are spun off from one another, and they need to switch tools or systems and start using a different platform. When customers use this platform, it becomes even direr that the transition is seamless. This situation is a perfect fit for the MOC. Starting from one platform, maybe it's a communication platform like a ticketing system for a support organization and a separate think tank/online forum for customer recommendations. This is our initial state, then consider, for example, the future state being a situation where these two platforms are consolidated into one. It could be for cost-saving purposes or a consolidation effort to make the customer experience easier to manage.

The future state is different from the current state but not radically impossible. For example, suppose system settings or APIs cannot be utilized. In that case, the transition might be difficult, and a negative sign would be listed in the MOC, indicating conflicting systems that are not easily merged. What about the customer data that needs to be transitioned? If what is in the support portal and what is in the communication forum easily compatible with the new tool, then this transition is simple, and a positive sign can be included to indicate this. So far, we are 1:1 when it comes to negative vs. positive. How about the number of personnel required to manage this tool? Let's consider four people for each of them in the initial state, and in the future, there will be five employees managing the consolidated tool. This one may be harder to gauge because there will be three fewer individuals managing the same data, but it will all be done within a single platform rather than two separate ones. A question mark would be placed in the MOC, connecting these two states to indicate uncertainty until more data becomes apparent. Another important area in this exercise is whether all features from the original two tools will remain in the new tool. For the sake of this model, we will consider that 4 out of the 25 primary features are missing. As a result, the future state would require workarounds to be developed through other existing features or some extra steps to deliver the same results. If no workaround is there, can the users live without this feature? Will it completely diminish certain aspects of the customer experience? Will a customer-wide public announcement need to explain the changes to customers and the impact of not having certain features readily available when the transition happens? If this is the case, a negative sign will be placed in the center matrix connecting the two states.

To summarize, these are the results of the center matrix for the transition across these two states.

- Consolidation of two software platforms into one with complex settings: -
- Migration of data from two databases and consolidating into a single, compatible database: +
- Reduction of the number of employees managing the two separate tools when the migration happens to a single too:?
- Lack of available to support some major features in the new tool: -

Out of the four items being modeled across the two states, two are negative, one is positive, and one cannot be determined. So we see that this will not be an easy transition. But that doesn't mean it is impossible, further testing may need to be conducted, and a deeper model may need to be fleshed out in which we create sub-categories for the items we are examining. We know that running this transition through the MOC shows us that a certain number of conversations need to occur amongst the stakeholders to find ways to make the change simpler.

Continuous improvement is one area we think is very hard to model within the MOC. This is because continuous improvement is often (not always) an exercise focusing on what the organization could be doing right and identifying any shortcomings in an already very useful concept.

It's a tricky situation because what if the change implemented by an organization ends up breaking something that was already functioning well? The MOC as an exercise in this situation should probably be superseded by other prioritization and risk exercises to determine whether the functions being examined truly need to be modeled. What is the business's risk if this change's outcome is negative? A Risk Assessment Matrix is a great tool for this situation (Usmani, 2022). Figure 3.7 provides a good example of a Risk Assessment Matrix.

Nobody wants to see a great functioning system fall apart because employees miscalculated an opportunity to improve an area that may not have needed any improvements. In contrast, accepting the state and never revisiting your systems following a certain cadence is extremely bad practice. There is always something that can be improved. The hard part is determining what the impact could be if this decision turns out to be a mistake.

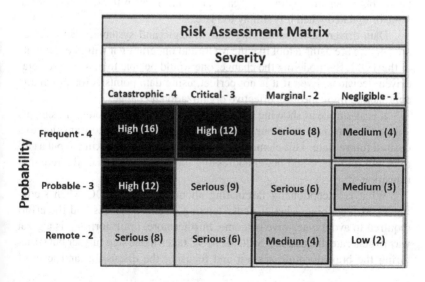

Risk Assessment Matrix			
Severity			
Catastrophic - 4	Critical - 3	Marginal - 2	Negligible - 1
High (16)	High (12)	Serious (8)	Medium (4)
High (12)	Serious (9)	Serious (6)	Medium (3)
Serious (8)	Serious (6)	Medium (4)	Low (2)

Probability: Frequent - 4, Probable - 3, Remote - 2

FIGURE 3.7 Risk assessment matrix

Moving from something that could be very hard to model in the MOC to something that makes perfect sense, we look at the next key characteristic of the Total Quality Matrix: Effective Communication. More often than not, an area within large organizations needs to be improved. But, unfortunately, one of the biggest drawbacks of growing businesses is growing Silos. And these types of separations break down progress and impactful results.

Every team should know all the cross-functional individuals they need to work with. If they are not collaborating, the organization needs to determine why. With the MOC, we look at the transitions needed to make more effective communication a feasible objective. But difficulties could arise in several ways and be something as simple as geographic location. If two teams that need to work together are sometimes stationed worldwide, one in the Mid-West US and the other in Hong Kong, that could make ongoing collaboration quite difficult. It's not impossible, especially in today's age. Several of us may have 9 pm calls scheduled to cater to a colleague overseas. It works, but it may not be ideal, opening up other areas for review within the MOC. Should certain teams be moved? Should everyone on the team in the Mid-West US be let go and a mirror team established in Hong Kong for easier communication? Our original MOC developed to improve communication is now opening up a second MOC to determine whether a new team should be formed worldwide. And again, this could be another scenario for additional tools to support the exercise.

But geographic conflicts are just one simple situation we've chosen to model here. Other more complex ones can include competing priorities where teams focus on engaging with a certain area of business to move initiatives along, but other groups in the organization are kept in the dark, out of the loop, or updated when it is simply too late.

Data-driven decision-making and strategic and systematic approaches are why we are using a tool like the MOC and integrating it with the elements of the TQM. By modeling the change, one would be able to make these data-driven decisions. Even if it is not perfect, using data points is the pragmatic approach to implementing effective organizational change.

It is as simple as showing the leadership team that X amount of elements within the current state of our organization will be difficult to transition to a desired future state. This element provides them with the backing to put a call to action into place and begin addressing the toughest parts of the required transition.

What is exceptionally fascinating about using the MOC when studying TQM is gaining a better understanding of consequences and the effort required to avoid a negative outcome. Furthermore, brainstorming is a great way to generate discussions. Still, the MOC takes it one step further but structuring the brainstorming session and focusing the discussion and level of brainstorming. It allows all stakeholders to zero in on the most critical areas that must be looked at closely.

Adopting a TQM approach is an excellent way to ensure your organization performs at the highest level, but getting there doesn't just happen overnight. Trying to force change without a good sense of how the business can be impacted could be very frustrating.

THE HOUSE OF QUALITY

There is a relationship between the two modeling techniques we are outlining in this book which goes back to their origin. The House of Quality is a prerequisite for TQM, and the MOC is derived from – the primary modeling tool at the core of *Quality Function Deployment* (QFD). This management concept has been so expansively adopted by some of the biggest organizations in the world (Hauser & Clausing, 1988).

Although not the focus of this book, it's important to understand how the tools we discuss here came to be. Like the MOC, the House of Quality examines the interactions amongst different states and business areas, tying together a full picture behind an operation rather than a siloed view. One of its main advantages is that the House of Quality helps minimize confusion and misunderstanding. It is an extremely efficient way to translate customer needs into actionable organizational items. The model is built around the *Voice of the Customer.*

From an engineer's standpoint, customer needs are always our guiding principles, which dictate (or should dictate) how we operate. This is why these are key components within the House of Quality. These customer needs are benchmarked and prioritized within the House of Quality to compare what the customer is asking for to the technical requirements outlined by the organization. While building the House of Quality, the customer needs are compared to how the organization understands these requirements, how it delivers on them, and how an organization may believe competitors are doing to address those customer needs.

Once an organization outlines what they believe the technical requirements are or translates the customer needs into actionable items, they then compare each of those requirements to one another to understand how they interact (or conflict). This process is very similar to the MOC (see Figure 3.8). In simple terms, this is great for ensuring customer goals are not getting in the way of each other.

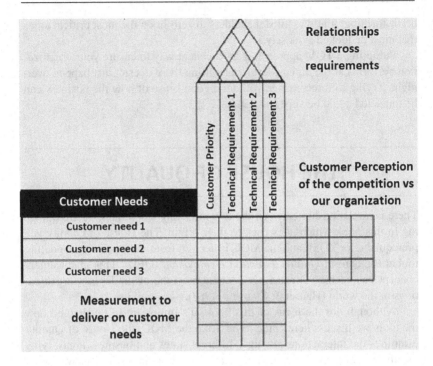

FIGURE 3.8   The foundation of the house of quality

As mentioned in this book, organizations fail when they look at a problem from a single angle. Trying to deliver on a customer's requirement without understanding its impact on other critical requirements can create gaps and bottlenecks or even open the door to a domino effect of problems that may not have been visible on the surface. The MOC works to identify relationships amongst existing operations, as well as future operations. The House of Quality works to identify relationships amongst a voiced need by the customer. It outlines requirements by an organization and the execution against them by both an organization and its competitors.

To go into the details of how the House of Quality fully addresses these factors would require its book, but this brief explanation is meant to provide a glimpse into the connection between the different tools, how they draw inspiration from one another, and how they can even be used together to give a holistic view of the way an organization can improve and aid the leadership team in making productive decisions

QUALITY 4.0

The Internet of Things (IoT), Industry 4.0, and Digital Transformation/Digital Thread have brought Quality 4.0. According to Escobar et al. (2021), Quality 4.0 is the "fourth wave in the quality movement (1. Statistical Quality Control, 2. Total Quality Management, 3. Six sigma, 4. Quality 4.0)." Quality 4.0 builds on the previous philosophies/frameworks. However, it takes advantage of the big data generated by IoT and the analytic capabilities of Artificial Intelligence to provide new solutions to problems that are hard to solve.

Quality 4.0 will still use the tools of the previous Quality philosophies/frameworks. Therefore, we will include this in the future frameworks for leader selection which will be more data-driven with more behavioral/cognitive analysis, which Artificial Intelligence will provide.

Case Study
Yahoo!

4

Case scenarios are an appropriate method to see the framework's capabilities being covered in this book and, by default, improve upon it. For example, one very functional case study is the very well-known case of Yahoo!'s hiring of Marissa Mayer as CEO (Elattar, 2014). At one point, Yahoo! was on a downward spiral, had gone through several CEOs in a very short time, and was faced with a fast-growing giant of competitors. As a result, Yahoo!'s scenario was viewed as a recovery effort. However, this scenario to be discussed further was from 2010 to 2014, and the year of the decision-making application was 2012.

Yahoo!'s case is an interesting model of a company so lost it cycled through five CEOs in only two years (2010–2012). Finally, things got so bad that the company had no choice but to hire one of the executives from their primary competition to set things right. Yahoo! was, at one point, the Google of the world. It was groundbreaking, it was efficient, and it was easy to use. But in such a competitive field, Yahoo! failed to adapt adequately to the mobile revolution, the ongoing demands of consumer cloud storage, and their dire need for more accurate search results. The ship was not sunk, but its path constantly changed in the wrong direction, and it needed swift action to become competitive again.

This case study looks at Marissa Mayer's introduction to Yahoo!. This example emphasizes a struggling company needing to change to recover from a troubling downward spiral. We specifically utilize this example to observe how a Matrix of Change (MOC) and a leadership classification scheme can be used to help organizations better understand their current situation and where they can be in the future if they take the proper actions.

YAHOO! CASE STUDY DESCRIPTION IN 2012 (ELATTAR, 2014)

Yahoo! was a well-recognized brand worldwide. Its products and services enabled the company to attract, retain and engage users, advertisers, publishers, and developers. As a result, Yahoo! positioned its products and services as the center of the world's digital daily habits. In addition, the company provided communications tools to connect the world and User-Generated Content products. Many of its properties were also available in mobile-optimized versions for display on mobile phones and tablet devices or as native applications for iOS, Android, and Windows phones.

As one of the Web's largest publishers and the owner of leading properties across multiple content categories, Yahoo! provided contextually relevant content and experiences where advertisers could connect with users effectively. It also brought quality publishers together, like AT&T, Verizon, Rogers, Monster, and Comcast. In addition, agreements with Microsoft and AOL allowed ad networks to offer each other's premium, non-guaranteed online display inventory to their respective advertising customers.

Yahoo! continually launched, improved, and scaled products and features to meet evolving users, advertisers, publishers, and developers' needs. Most of the software products and features were developed internally. Yahoo!'s employees and culture were fundamental to the company's success, and attracting the best people to Yahoo! was critical. It included a broad array of engineering talent that spans the breadth of technology infrastructure, primarily located in California, India, and China.

Finally, Yahoo! had the capital available for further growth. In September 2012, Alibaba Group Holding Limited repurchased 523 million ordinary shares of Alibaba Group owned by Yahoo!, resulting in a pre-tax gain of approximately $4.6 billion for Yahoo!.

On the other hand, Yahoo! main weaknesses were its inability to defend its market position in the search marketplace and its unstable management during the last five years (2007–2012). In addition, the development of Yahoo!'s search engine technology could be described as inconsistent, hampering the enhancement of its search engine capabilities. In the beginning, Yahoo! began using Google's search technologies to exploit the booming Internet fully. By 2004, it had changed back to its search technologies. However, the further effort to compete with Google in the search market with the Panama project in 2006 did not give the expected results, and Google maintained its superiority. Then, the Search Deal with Microsoft made Microsoft Yahoo!'s provider of search technology and advertising. This arrangement made Microsoft the #2 search engine behind Google, gaining Yahoo!'s market share.

Adding this to the unstable period in the executive board, where there have been around five to seven CEOs in the last five years, Yahoo! had shown weak performance compared to the competition, reporting stagnant revenues. The last weakness was the inappropriate handling of Asian partnerships, especially with Alibaba, which showed poor leadership and a lack of diplomacy.

Scott Thompson was the CEO at Yahoo! from January 2012 until mid-May 2012, a very short tenure in which Thompson initiated a 14% reduction in the company's workforce (Efrati & Bensinger, 2012). This decision did not sit well with several executives, who resigned before the layoffs started. However, his 130-day tenure gave him over $7 million in compensation.

Thompson was temporarily replaced by Yahoo! Vice President Ross Levinsohn, who served as interim CEO until Marissa Mayer took over the position in July 2012 (Efrati & Letzing, 2012).

PRELIMINARY ANALYSIS (ELATTAR, 2014)

The Five Forces of Porter provides a good framework to study and understand the competitive environment. For this study, we began by examining Yahoo!'s advantages by using Porter's five competitive forces (Figure 4.1).

Barriers to entry were high and steady. New entrants were generally startups with new technologies. They required access to skilled human resources and substantial computing resources. The dominance and strong branding of incumbents characterize the industry. However, opportunities exist for those with a narrow focus on the new applications of search technology or on developing new technologies that enhance existing functionality (location-based services, users' intent). Industry giants usually acquire these new entrants.

Competition is high, and the trend is increasing. Concentration increased steadily, primarily driven by Google's growth and the decline of smaller search engines (Ask.com and AOL). As a result, the market share of the search engine industry is distributed in the following way: Google (77.3%), Microsoft (9.0%), and Yahoo! (8.8%).

The competitive factors for both user and advertiser levels are described in Table 4.1.

NEW ENTRANTS

- Start-ups with new technologies
  - location-based services
  - Users' intent
- Non-US companies

SUPPLIERS

- Internet publishing and broadcasting
- Internet service providers
- Advertisers
- Mobile companies

COMPETITION

SEARCH ENGINE INDUSTRY
Google: 77.3%
Microsoft: 9.0%
Yahoo: 8.8%

CUSTOMERS

- Retail trade
  - Retail: 23%
  - Telecommunications: 14%
  - Financial services: 13%
  - Automotive: 11%
- Advertisers
- General users

SUBSTITUTES

- Social media networks
- Potential for a new product
- Conventional advertising

FIGURE 4.1   Porter's five competitive forces as applied to Yahoo! (Early 2012)

TABLE 4.1   Yahoo!'s Competitive Factors

USERS	ADVERTISERS
Relevant search results	Size of user base
Aesthetics	Cost per click charged
Speed of service	Number of affiliated websites
Branding	
Free services, software, or online storage	

Globalization in the online services industry was on a medium scale in 2012, increasing the trend. For example, all US companies had 50% of their revenue outside the US, whereas non-US companies were entering the US market. On the other hand, the regulation level was light but expected to increase.

Substitutes also tended to lie on a medium scale, increasing the trend. However, increased external competition from social media networks was challenging the industry. Therefore, integrating search engines and social networks enhanced users' search experiences. Some examples of this "social search" were Google+ with its search engine and Microsoft with Facebook.

But the bargaining power of customers remained high. Moreover, the availability of several alternatives at no cost for the customers and undifferentiated services gave customers more room for negotiations and for choosing which direction they needed in the future.

Bargaining power for the suppliers was on a medium scale. This situation was because numerous suppliers, programmers, and advertisers could work with different customers simultaneously. However, the preference to do business with the industry giants (Google or Microsoft) negatively affected suppliers' bargaining power.

After the Five Forces of Porter, a SWOT analysis can be built to understand the capabilities and potential for future directions (Figure 4.2). For

STRENGTHS	WEAKNESSES
• A well-recognized brand around the world • Strong focus on distribution • Distributes its content across several screens • Remains one of the most visited sites • Alliances with leading players • Established distribution partnerships • Increased multi-platform offerings and social network integration • Cater to multiple audiences • Exploit the emerging trends of social networking effectively • Capital available to invest aggressively • A highly skilled workforce	• Inconsistency in developing its own technology in the search engine • Weak performance compared to peers/stagnant revenues • Inability to defend its market position in the search marketplace • Unstable period: 5 – 7 CEOs (Chief Executive Officer) with 5 years • Partnership with Microsoft • Microsoft surpasses Yahoo as the second giant in the industry • Lost or deterioration of Yahoo's search engine capabilities • Strained relationships with Asian Partners show poor leadership from the board and a lack of diplomacy
OPPORTUNITIES	THREATS
• Shift of consumers to mobile devices • Online advertisement market is one of the fastest-growing segments • Sustainable growth • Estimated growth of 31% in 2014 • Display advertising and mobile ad spending will be strong emerging drivers • Display ad 2012-2014 growth: 47% • Display ad 2012-2014 growth: 85.4% • Rapid Technology and process change	• Intense competition in online display ad marketing from several sources • Facebook is more cost-effective for advertisers • Sustainable competitive advantages are not assured • Google and Microsoft have launched several initiatives to capture share in these markets • Applicants like Siri for iPhone have introduced users to reality without search engines • Non-US companies entering the US Market • Increased dependence on advertising revenues exposes Yahoo to business cycles • Unpredictable and variable, more vulnerable and higher risk • Online fraud is becoming a significant issue

FIGURE 4.2 Yahoo!'s SWOT analysis (Early 2012)

example, to understand Yahoo!'s standing in the market, we developed a SWOT analysis to see what factors the company should consider when organizing its management and building well-backed strategies.

The opportunities for Yahoo! arose from the shift of consumers to mobile devices and the fast growth pace in the online advertisement market in 2012. Therefore, it was expected that display advertising and mobile ad spending would be very strong emerging drivers within this industry.

The main threats were strong competition in online display ad marketing from several sources and increased dependence on advertising revenues that exposed the company to vicious business cycles. This dependence would make Yahoo! more vulnerable and exposed to higher risks. Another issue that was gaining importance was the growing concern of online fraud, which is where companies would invest to ensure their credibility in the market.

YAHOO'S MATRIX OF CHANGE (ELATTAR, 2014)

We developed a MOC based on Yahoo!'s situation to expand on this study. Therefore, we used the information provided by Porter's Five Competitive Forces and the SWOT Analysis to build the Current State (i.e., existing practices) of the MOC for Yahoo!. The Current State has four main areas: Operations & Leadership, Human Resources, Growth Capacity, and Innovation & Product Development (Figure 4.3).

In the Operations & Leadership area, the practices were:

- One of the most visited and trusted websites
- Content, communication, and community platform
- Operations mainly in the US
- Eroded confidence in the board
- Inconsistent regional strategies

Yahoo!'s Human Resources were characterized by the following:

- Highly talented workforce
- Low employee morale
- Work-from-home arrangements
- Employee Stock Purchase Plan

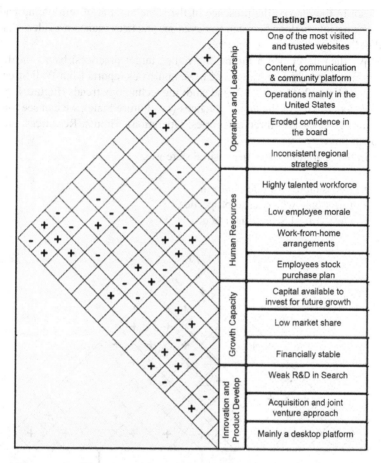

FIGURE 4.3 Yahoo!'s existing practices

The Growth Capacity area could be described by:

- Capital available to invest in future growth
- Low market share
- Financially stable

The Innovation & Product Development practices were:

- Weak R&D in Search
- Acquisition and joint-venture approach
- Mainly a desktop platform

As seen in the matrix, the presence of the same amount of reinforcing and interfering interactions between the existing practices makes it evident that Yahoo!'s Current State was unstable.

We then developed a Future State (i.e., target practices) based on the interviews conducted by Yahoo!'s CEOs, numerous reports from Wall Street analysts (Years 2011 and 2012) and studying technology trends (Figure 4.4).

When looking at the target practices (i.e., Future State), we can see that they are grouped into three categories: Operations, Human Resources, and Growth Strategy.

In the operations area, the desired state was:

- Focus on the Mobile platform
- Offer only widely used properties
- Customized Search Interface

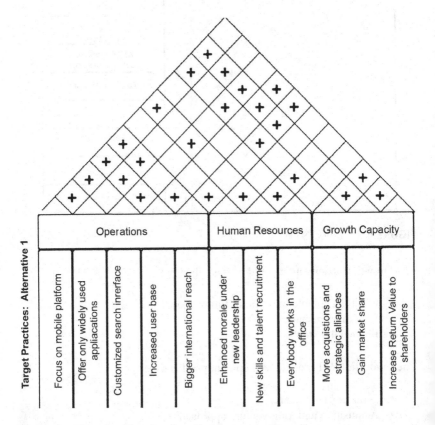

FIGURE 4.4 Yahoo!'s target practices

- Increased User Base
- Bigger International Reach

The company's goal was to have 50% of operations focused on Mobile applications. Yahoo! already worked in small, fast project teams focused on iOS and Android platforms. Also, Yahoo! was selling ad space for their mobile applications ahead of time and searching for ways to monetize mobile properties quickly. The goal was to offer only widely used properties, their focus being every 'digital daily habit, ' and the executive team gave the 'go-ahead' only to projects that could scale 100 million users. Yahoo!'s core priority was to customize its search interface. By innovating at the user level, searches were personalized. The engine should know where you are, the context where you are searching, and who you are communicating with. In addition, there were planned improvements in voice recognition, image recognition, and translation. All these efforts were going to increase the user base. This increase was another main goal of Yahoo! and achieving a bigger international reach since 75% of Yahoo!'s revenue came in 2012 from the America Region. There was a great opportunity to increase its market share as it expanded.

The goals for the Human Resources area were to enhance morale under the new leadership, develop new skills, recruit top talents, and put everybody to work in the office. The latter would enhance a collaborative culture and improve employees' productivity.

For the Growth Capacity area, the targeted practices were:

- More acquisitions and strategic alliances
- Gain market share
- Increase Return Value to shareholders

Yahoo! aimed to be a more global engine that users touch daily; its executive team should continually innovate to increase customer satisfaction and, therefore, gain lost Market Share, increase Revenue and Increase Return value to stakeholders. In addition, Yahoo!'s strategy should also involve making key acquisitions and alliances to be competitive. Therefore, partnerships were key to their future success.

The dominance of positive interactions between these target practices makes the Future State of Yahoo! stable.

Next, we will look at the transition matrix, which helps determine the degree of difficulty in moving from the Current State to the Future State (Figure 4.5). The proportion of positive and negative signs in the transition matrix indicates how disruptive the change process will be. A transition matrix with a comparatively large number of complementary practices and a few conflicting practices indicates that the change will be relatively

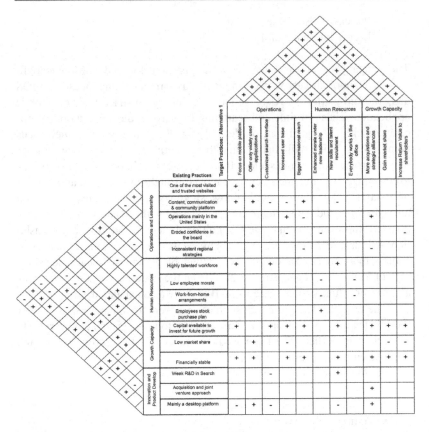

FIGURE 4.5 Yahoo!'s completed matrix of change

incremental and non-disruptive. In the case of Yahoo!, the transition matrix has slightly more reinforcing than interfering interactions. This type of matrix means that the transition will not be so smooth. This transition matrix indicates that the change must be made in phases.

CASE STUDY FINDINGS
AND ANALYSIS

When Google came into the picture, it went head-to-head with Yahoo! and came out on top from several angles. Google provided a faster and more accurate search engine while adding some features. It developed its email system, providing a much larger allowance concerning attachments and actual mail storage. The icing on the cake was the launch of their custom web browser and operating system, which grew to be the number-one mobile platform in the world in a matter of years. On the other hand, Yahoo! was quite limited in resources and could not be competitive; the endless cycles of CEOs did nothing but cause disruptions and lower employee morale. These chief executives provided various opinions, but none of them could look at the heart of the problem.

One major advantage Marissa Mayer had as she came into Yahoo! came from Yahoo!'s chief competition. In other words, she knew exactly why Yahoo! was failing and why Google was consistently overtaking them. As a result, Marissa Mayer did not shy away from taking action. *Directing* Leadership is the key for companies that need a series of projects complemented by acquisitions. Marissa Mayer had the best *Directing* experience at Google and participated in major projects such as developing Gmail and Google Maps. Yahoo!'s major pitfall at the end of the day was a lack of productivity; this stemmed from many factors, including the organization's work-from-home policy, lack of presence in major international markets, and a dire need for a more diversified portfolio. Instead, Mayer focused on the company's strengths and added more fuel to the Flickr division through its expensive acquisition of Tumblr. In addition, she improved the aesthetics of the company's main website and mail system. These actions were financially proven to be the right decisions in a very short time. Therefore, the *Directing* Style should be the dominant style in the first two/three years. However, after that, the style must evolve more balanced with *Cultivating* and *Visionary* to become stronger to grow and innovate continuously. *Cultivating* will keep sustainability by supporting a good working environment.

Therefore, the first two years of *Directing* (and Marissa Mayer was capable of this) is the key. *Directing* during the first two years can stabilize Yahoo!. But the challenge is to build a more balanced style after the first two years to compete against Google and Microsoft or to achieve a higher stock level to sell Yahoo! with a higher value to satisfy the shareholders.

Yahoo!'s case study was a perfect example of utilizing the MOC. It is a tool that would have been very beneficial for Yahoo!'s board of directors to use themselves to understand where their company needed to go. Even though this research has repeatedly stated that the MOC is not a conclusive tool, it is only an initial visualization for corporate positions. Our application of the tool in Yahoo!'s example illustrates that no conclusions can be made based solely on the MOC; it is only a first step to giving stakeholders a better idea of implementing certain changes.

Instead of going through multiple CEOs giving no results, utilizing the MOC would have given them a better idea of the right person needed to take the company in the right direction. Marissa Mayer may not have been available when Yahoo! had gone through the first of five CEO changes. Still, it would have given the board insight into the company's mistake by choosing any of them.

CONCLUSIONS

In conclusion, the Yahoo! case study was a great way to assess the framework discussed in this book. Yahoo! leadership case study shows how well the techniques work in practice. Worldwide, people are familiar with Yahoo!. With the help of its goods, services, and content, the business could draw in, keep, and engage users. Aside from being up against a huge competitor that was rapidly growing, Yahoo! was in a downward spiral and had gone through several CEOs in a short time. However, Yahoo!'s strategy was viewed as a recovery effort by many, and the *Directing* style was good for regaining positions.

To begin this assessment, we first look at Yahoo's advantages using Porter's five competitive forces. Then, to further this study, we decided to build a MOC based on Yahoo!'s specific situation. Next, we used the information that Porter's Five Competitive Forces and the SWOT Analysis provided as we began to build Yahoo!'s Current State of the MOC. Finally, the transition matrix was examined to determine how difficult it would be to move from the Current State to the Future state. This examination indicates that the transition matrix for Yahoo! features slightly more reinforcing interactions than interfering interactions. These interactions suggest that the transition was not going to be as smooth and that different phases were required. The MOC can help us see how the leadership styles map to different phases and the respective candidate.

The Yahoo! case study served as the ideal illustration of how to apply the MOC. Using the MOC would have given them a better notion of the proper leader needed to lead the company in the right direction rather than passing through numerous CEOs with no results.

Advances in Leadership Development

5

Due to the importance of leadership development, enterprises spend massive amounts of money on training their leaders. Leadership development, in general, went through many phases and practices to satisfy the need for leaders who can make a difference. Virtual Reality/Virtual Simulation can create cost-effective leadership development environments.

The advantage of using Virtual Reality/Virtual Simulation in Leadership is that it is great in transferring the complexity of real-life scenarios to controllable systems. In addition, computer-aided design (CAD) tools replace physical world elements with a computer-generated or virtual environment (VE) or virtual world (Almalki, 2016).

The most effective method of training and development is experiential learning. In addition to real-life experiential learning, simulation offers a viable alternative to the traditional face-to-face (classroom/textbook/lecture) mode of instruction used in many schools and universities. Virtual environment simulation-based education not only complements but also enhances traditional teaching methods. It can also capture users by generating a sense of being there in an area, bridging the gap between content understanding and experience learning (Bhide et al., 2015; Lemheney et al., 2016). In addition, Virtual Reality/Virtual Simulation is the core of making the Metaverse functions and can be utilized to build digital twins.

FIRST-GENERATION EXPERIMENTS

Our first generation of Virtual Simulation experiments was based on the 4-D Leadership Classification System (explained in Chapter 2) and Virtual Worlds built using OpenSim (Allison et al., 2012). The experiments were

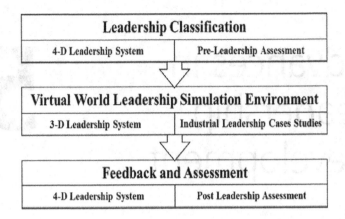

FIGURE 5.1   Framework for leadership development using virtual worlds

based on comprehensive case studies, as shown in Figure 5.1, which were experimentally examined to prove the effectiveness of virtual simulation in leadership development. Our first experiments use engineering subjects.

In a virtual world, the real world can be recreated in a three-dimensional, computer-based, immersive setting, like the 3D virtual world of Second Life, over the internet (Hudson et al., 2014. Virtual Reality goes by many names, including artificial Reality, virtual worlds, multi-user virtual environments (MUVEs), massively multiplayer online game games (MMOGs), and immersive virtual worlds (Bamodu & Ye, 2013; Girvan, 2018).

Virtual world simulation was chosen for this study because it has been successfully used as a tool for training, simulation, and teaching (Sequeira & Morgado, 2013). Virtual simulation can give users shared, concurrent experiences and opportunities for conversation. It motivates leaders to take more chances and explore while learning novel concepts and methods (Siewiorek et al., 2012). Trainees can benefit greatly from 3D virtual worlds in their learning process. They can safely make mistakes in simulated environments and learn from them, gaining the same experiences as they would in a real-world scenario without the challenges posed by cost, time, and ethical considerations (Williams-Bell et al., 2014).

The junction of three environments – the visual world, the auditory environment, and the haptic/kinesthetic environment – creates the virtual environment. This synthetic sensory experience conveys abstract elements to participants (Çapin et al., 1999), which is important for developing environments to develop leadership skills. The components of a virtual reality system can be found below (see Figure 5.2). The virtual reality engine and input/output (I/O) devices are hardware components; application software

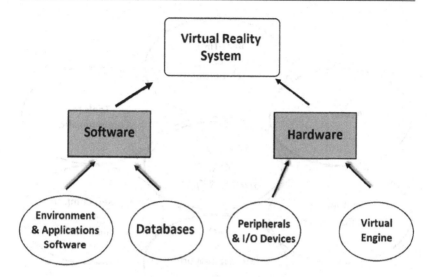

FIGURE 5.2 Components of virtual reality systems

and databases are examples of software components (Bamodu & Ye, 2013). The input devices provide users with methods for interacting with the virtual environment. The output devices stimulate the users' senses, which take feedback from the virtual reality engine and deliver it. Therefore, a computer system called a virtual reality engine needs to be chosen based on the user, I/O devices, graphic display and image production, and level of immersion (Bamodu & Ye, 2013). The equipment selection was straightforward for our experiments which involved OpenSim to develop the environment, a server, and computers/laptops for the user to access the Virtual World.

The tools and software used to create, manage, and maintain virtual worlds, as well as the database, are collectively known as virtual reality system software (Bamodu & Ye, 2013). Figure 5.3 below shows the resources of the virtual environment simulation. In addition, our team got specialized support from experts in OpenSim, multimedia, usability, and aesthetics of virtual worlds.

The 3D virtual world for Leadership is a simulation where the real world can be recreated in a three-dimensional, computer-based, immersive environment like Second Life which allows interactions with "live" agents (Elattar, 2014; Hudson et al., 2014). The term "virtual world" refers to a simulated 3D virtual world environment that offers chances for simultaneous interaction between users and the 3D simulated environment. Along with other tools like text-based or voice-based discussions or the ability to roam between several simulated geographic locations, this is done through

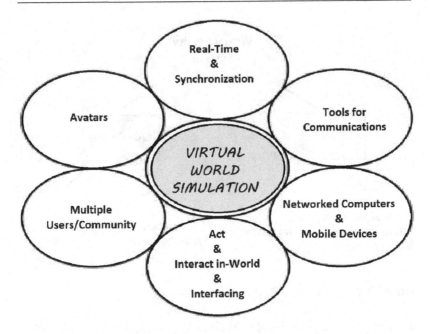

FIGURE 5.3   Virtual world simulation resources

a user's avatar. Users can express themselves and have mobility between various simulated geographic places. This happens through a user's avatar (Cruz-Benito et al., 2015). Users access the 3D virtual world simulation on the web using a graphic representation of an avatar that can fly, drive a car, walk about, and teleport to different environments to engage in various activities. These 3D virtual simulation qualities need the virtual environment to be created per the simulation's goals. That was the environment built for our experiments using OpenSim, where the user can work with other users (even some avatars are not humans but utilize Artificial Intelligence). Figure 5.4 shows Avatar selection and customization from one of the virtual worlds used in our research.

Figure 5.5 shows a building and a house built by teams in the virtual worlds. The experiment used several teams to build a city composed of houses following green, ecological, and sustainability guidelines. The experiment using the virtual world simulation and the traditional setting emphasized *Inclusive, Directing,* and *Visionary* Leadership styles.

The experiment (Almalki, 2016) used a total of 84 engineering students (statistically meaningful size) with very similar backgrounds. They were randomly invited: 42 in the experimental group (trained using the Virtual World) and 42 in the control group taught by using Microsoft PowerPoint in a

FIGURE 5.4 Example of avatars used in our virtual worlds

FIGURE 5.5 Examples of a building and a house built by teams in our virtual worlds

traditional setting (classroom)). In this experiment, both groups (experimental and control) were given the same pretest. After the pretest, the experimental group was subjected to the experimental treatment, which is a 3D virtual leadership simulation. The control group, in contrast, receives traditional engineering leadership development using the standard method of conducting engineering leadership case studies as a team: case study analysis lectures, reports, and leadership solution presentations. Therefore, this group is completely cut off from the 3D virtual world simulation treatment. The experiment for both groups took several weeks to complete.

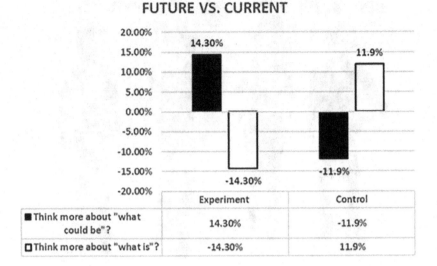

FUTURE VS. CURRENT

	Experiment	Control
■ Think more about "what could be"?	14.30%	-11.9%
□ Think more about "what is"?	-14.30%	11.9%

FIGURE 5.6   Thinking about the future or current matters

The results of this experiment revealed great insights regarding using 3D virtual world simulation in leadership development. For example, the 3D virtual world simulation participants preferred considering a futuristic thinking approach over the current situation thinking approach by almost 14.3%. In contrast, the control group participants shifted by almost 12% toward thinking about the organization's current situation (see Figure 5.6). This indicates that the 3D virtual simulation helps engineering students develop imagination and visualization of the future (a strong component of *Visionary* Leadership).

Other results support that the 3D virtual simulation enhanced the *Visionary* and *Directing* leadership styles more than the study's traditional method, providing the students with a more balanced style (see Figure 5.7).

Decisions are made with consensus in mind. Figure 5.8 reveals an interesting outcome, demonstrating how the 3D virtual world leadership simulation had an unequaled and exceptional impact on team development in this study. When using the traditional method, there was only a 4.8% shift in the control group, whereas there was a 14.5% move in the experimental group in favor of seeking agreement. This finding is consistent with the idea that the Virtual World may impact followers' capacity for teamwork.

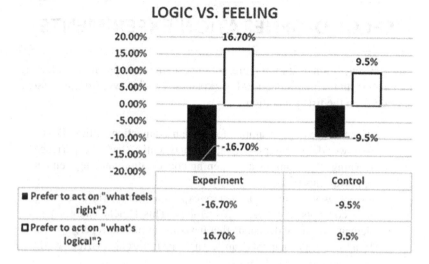

LOGIC VS. FEELING

	Experiment	Control
■ Prefer to act on "what feels right"?	-16.70%	-9.5%
☐ Prefer to act on "what's logical"?	16.70%	9.5%

FIGURE 5.7   Prefer to act with logic or feelings

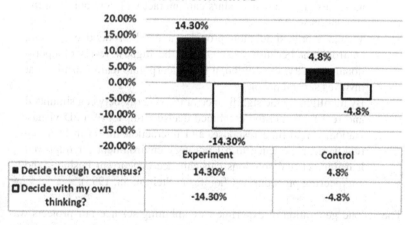

DECIDE THROUGH CONSENSUS VS. THROUGH OWN THINKING

	Experiment	Control
■ Decide through consensus?	14.30%	4.8%
☐ Decide with my own thinking?	-14.30%	-4.8%

FIGURE 5.8   Decisions based on consensus vs. own thinking

SECOND-GENERATION EXPERIMENTS

The first generation of studies represents a positive development in evaluating Leadership and attitudes in STEM workplaces. The following are the lessons learned from the first generation:

- It encourages opportunities for human connection in the 3D virtual world, giving participants concurrently shared experience learning opportunities and opening conversation and engagement opportunities.
- Virtual world simulation fosters experimentation while teaching new concepts, abilities, and methods. This benefit enables participants to make decisions that have no significant repercussions. In addition, participants can learn actively through virtual world simulation instead of passively.
- Virtual simulations allow participants to use their graphical representation, an avatar who can walk, fly, and operate a vehicle while simulating the real world in an immersive setting.
- Virtual Worlds have been utilized in training for the most advanced nuclear power systems, software engineering, military training, and aviation technologies. Without any preparation from a scenario designer, human avatars can interact with any object in the simulation world.
- Virtual world simulation provides a good productive learning environment because it can attain the highest levels of operational integrity. In addition, it may incorporate team learning that involves several people.
- By highlighting the significance of the synergy between simulated and real facts and the combined decisive impact of a 3D visualization, a real-time response, and interaction in 3D in bringing real-world experience to a laboratory environment for improved learning, virtual solutions can enhance concurrent business and engineering procedures (Chorafas & Steinmann, 1995).

The second generation of experiments combining science fair projects was performed to verify this investigation's validity and assess its potential with modifications from the first generation of experiments (Davis, 2020). In this second generation, engineering participants' leadership styles were evaluated before and after the experiment. Before the experiment, the 4-D Leadership System was used to understand the existing participants' leadership styles.

The leadership style was examined once more to determine the experiment's impact after partaking in the virtual world leadership experiment.

Every participant's percentage of each dimension was calculated (see Table 5.1). After the experiment, the average leadership percentage for each team member was determined. Since all leadership styles are geared toward the average score, the smaller the 4-D Leadership System divergence, the closer the Leadership is to being strong and well-rounded. As a result, the standard deviation measures how well-rounded and effective a group or individual's engineering is. It is known as the effective leadership indicator in this study (ELI).

To be included in the virtual world simulation, the research depended on the input and useful information from subject matter experts in Psychology and Education. The following criteria were used to choose the science fair projects:

1. It is engineering that combines technology and innovation.
2. It can be carried out in group situations.
3. It includes concepts from multiple engineering fields.
4. It makes sense, is easy to understand, and appeals to various engineering educational levels.
5. There is a room layout for creation and innovation in the virtual world.

To ensure that the leadership simulation in the investigation has the newest technological features and allows engineers to practice their Leadership with a complete, advanced, and fully functional platform, DreamLand Metaverse, the leading OpenSim software hosting service, has been used. In addition, we have strong support from experts in OpenSim, multimedia, usability, and aesthetics of virtual worlds. Figures 5.9 and 5.10 demonstrate the construction of the training environment in this project, which was built in a 250 × 256 square meter single OpenSim region measuring roughly 16 acres.

This area is an entirely virtual environment run by a single process. In addition, users can access other OpenSim grids through hypergrid capabilities, much like in any other virtual world that already exists. This mini-grid is hosted on a high-performance, dedicated, multi-core server tailored for OpenSim software as a fully managed service. The experiment was designed to develop more of the Directing Style of Leadership.

Twenty participants (with similar educational backgrounds) were chosen randomly to engage in the experiment. All 20 willingly participated in this experiment and were randomly divided into five groups. The five groups participating in this experiment received the same. Therefore, the experimental group will receive the experimental treatment, which consists of an immersive

TABLE 5.1 4-D and the Capacities of Teams and Individuals

LEADERSHIP STYLE	LEADERSHIP DIMENSIONS		TECHNOLOGICAL LEADER STRENGTH AND CAPACITIES OF TEAMS AND INDIVIDUALS
Visionary style	Information: Intuition →	Deciding: Logical	Logical and intuiting (also called the visioning dimension) depend on thinking about the future. Visionary leaders are influential leaders and usually initiate the things they want. Their specialties are invention, creativity, patent, and design.
Cultivating style	Information: Intuition →	Deciding: Emotional	Emotional and intuiting (also called cultivating dimension) promotes positive feelings and accomplishing a prosperous world, sincerely caring about each other and a sincere concern for humanity.
Directing style	Information: Sensing →	Deciding: Logical	Logical and sensing (also called directing dimension) stimulates taking charge and directing others. Planning, managing, directing, controlling, productive, and efficient are some actions of this type of Leadership.
Inclusive style	Information: Sensing →	Deciding: Emotional	Emotional and sensing (also noted as the including dimension) rely on the emotional capabilities that come from relationships and how we communicate with others – fostering the communication and relationships with others and conflict intolerant.

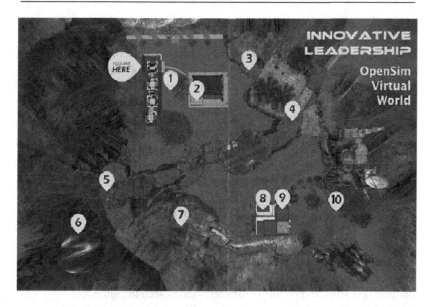

FIGURE 5.9   Top view of the virtual simulation area

INNOVATION & CREATIVITY
IN LEADERSHIP

The Research Experiment Path

1 Team Picnic

2 Innovation Theater

3 Science Fair Projects

4 Meditation Walk

5 Meditation Sphere Entry

6 Meditation Universe

7 Waterfall Overlook

8 Think Tank Brainstorm

9 Think Tank Inspiration

10 Team Collaboration Dropbox

FIGURE 5.10   Key to the virtual simulation area map

3D virtual leadership simulation, as opposed to the traditional engineering leadership development using the typical method of conducting engineering leadership case studies as a team, lecture analyses, reporting, and leadership solution presentations. In addition, the experimental group was exposed to the Virtual World Innovative Leadership Simulation Environment to determine whether their leadership abilities would have changed.

No prior understanding of virtual simulations or the virtual viewer program Firestorm Viewer was requested from the participants. As soon as the Virtual World environment was finished, the training began. Two weeks were allotted for the training. Participants underwent orientation training in the first week on using the Firestorm viewer and their avatars in OpenSim's 3D virtual environment. Participants were also provided instructions as homework assignments to become familiar with the viewer controls and 3D virtual world. They were using their avatars in the 3D Virtual World. Using the avatars in the 3D virtual world, each participant spent a minimum of three hours during the second week doing orientation exercises in a specialized virtual world setting that was used in conjunction with Zoom. All participants had their avatars personalized depending on their skin tones and physical characteristics, as seen in Figure 5.11.

Before the simulation, during the self-assessment, nine students were determined to be in *Directing* Leadership, five students were in *Cultivating* Leadership, two were in *Inclusive* Leadership, and four were *Visionaries*. However, the results indicated that students acquired more from the *Inclusive* and *Cultivating* Styles to complement their original styles.

FIGURE 5.11 Avatar selection and customization based on gender and skin color

TABLE 5.2   Results of 4-D Leadership Surveys (Pretest and Post-Test)

	PRETEST		POSTTEST		
	MEAN	ST DEV	MEAN	ST DEV	2 SAMPLE T-TEST
Emotional	4.8	1.48	4.4	0.98	$p = .32$ $t = 1.01$
Logical	4.4	1.04	3.9	0.88	$p = .11$ $t = 1.64$
Intuited	4.4	0.76	4.4	0.55	$p = 1.00$ $t = 0.00$
Sensed	4.2	0.63	4.9	1.08	$p = .23$ $t = -2.32$

Table 5.2 indicates the following results of the experiment:

- There was no significant difference in the scores for the Emotional Decider.
- There was a significant difference in the scores for Logical Decider.
- There was no significant difference in the scores for Intuited Information.
- There was a significant difference in the scores for Sensed Information.

Therefore, we can conclude that *Directing* Leadership was the dominant style of the participants in the Virtual World Experiment (however, the participants did not gain *Directing* attributes with the experiment). *Inclusive* and *Cultivating* gained with the experiment (reduction of Logical and increased of Sensed). The experiment was not conducive to gaining *Visionary* and *Directing* styles.

In addition, we analyzed the different tasks accomplished by the different teams. The results of these task(s) based on the rubric were proficient in research overview, organized preparation, and design creativity and originality. Teamwork problem-solving was determined based on the project task(s). The selection of these tasks, as mentioned before, was a result of the discussions with subject matter experts from Education and Psychology (See Figure 5.12). The task with lower points was related to design creativity and originality. Therefore, that justified that *Visionary* was not the dominant one for this experiment, and the subjects did not gain the respective attributes of *Visionary*. The results of the 4D tests indicated that too.

MEDITATION

Meditation is considered to be beneficial for controlling stress and helping visualization. This visualization can support the development of *Visionary*

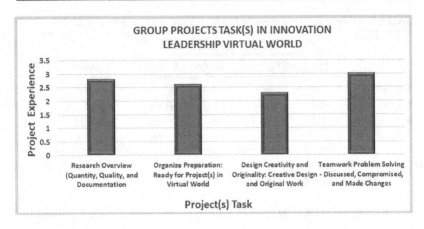

FIGURE 5.12   Assessment of Team Project Task(s) in the virtual world

Leadership. Birk (2020) discusses that Ray Dalio, the creator of the largest hedge fund in the world, lists meditation as the single most significant factor in his achievement. Another good example provided by Birk is the one of Mike Romoff. After meditating for several months, Mike Romoff, Linkedin's head of global agency sales, states that he has "had a gradual awareness that all beings are related, and the whole idea of having antagonistic thoughts towards others as autonomous entities ceased making any sense." Those are characteristics of the *Inclusive* style.

Technology, when used properly, can reduce stress. Such is the case with virtual reality meditation. Numerous peer-reviewed scientific research has shown that meditation is the best way to reduce stress (Allen, 2020). It can also help with medical disorders such as anxiety, chronic pain, high blood pressure, and tension headaches made worse by stress. As a result, more and more people have become aware of meditation and its advantages. Now, owing to technology, we can further improve the meditation experience through Virtual Reality. For example, Virtual Reality can transport you into your favorite video game, allowing you to run, fight, and explore your way through made-up landscapes, or it can enable you to visit museums all over the world from the comfort of your home. And with Virtual Reality, we can use Mediation to improve our Leadership and balance emotional and rational states.

Imagine that you can meditate in the virtual world. As part of this investigation, we added some components to the Virtual World of the second-generation experiments to perform meditation (see Figure 5.13). As a result, we had the opportunity to have the participants meditate while in the Virtual World environment. Initially, it seemed hard to imagine sitting on a fake

FIGURE 5.13   A place built in one of our virtual worlds to do a group meditation

beach watching a waterfall and mountains, but the participants were able to have this experience.

For instance, when using Virtual Reality/Virtual Simulation, the participant starts by gazing down a river. They can see trees on either side of the river, water ripples, and pebbles at the bottom of the river (Navarro-Haro et al., 2017). The participants can now see virtual mountains hidden when looking down the river if they move their heads to the left, and they can see boulders at the bottom of the river if they look down. Leaning forward and looking down, they can see the river getting closer.

After traversing through the Virtual World created for Leadership, the participants take advantage of finding their creative inspiration for innovation by entering the meditation sphere. After teleporting to the meditation sphere, the participants are provided instructions that will allow them to immerse themselves in space to imagine the possibilities.

Once in the Meditation sphere, participants can explore the limitlessness of their soul beyond their body, life, and self with guided meditation. When the participants move beyond their physical selves, they can begin to tap into infinite awareness, limitless knowledge, and their expanded, whole, energetic self.

We have all we need to be exceptional leaders. However, in certain situations and times, we could feel restrained or challenged and unable to act on our inspiration and vision. In this investigation, we are currently studying how Meditation in the Virtual World can help Leadership, particularly the *Inclusive* and *Visionary* styles.

As a result, in addition to using Virtual Reality to learn how to practice meditation, individuals may be able to practice meditation at work, home, or school and potentially improve their health and leadership capabilities, particularly the *Visionary* dimension.

METAVERSE AND DIGITAL TWINS

The metaverse is the "next iteration of the internet that seamlessly combines our digital and physical lives" (McKinsey, 2022). Therefore, the metaverse is a 3D model of the physical world where you spend your digital life. Currently, with Virtual Reality/Augmented Reality, we are experiencing the emerging (or "proto") metaverse. The metaverse has several initial features (McKinsey, 2022):

- a strong feeling of immersion
- real-time interactivity with extensive connectivity
- user agency with some level of Artificial Intelligence

And, when the most advanced versions of it are present, then other important features will be essential:

- interoperability - the ability to share and exchange information between multiple systems but also be able to work together to execute an operation or perform a complimentary function (synchronization in logic, geospatial (actual and virtual), and time)
- hyper concurrency and connectivity (real-time and hierarchically/ distributed)
- use cases covering human life (such as Leadership involving human behavior)

Therefore, Virtual Reality, Augmented Reality, and Virtual World Simulation are strong pillars of it. The metaverse can be used for Leadership development too.

On the other hand, the Digital Twin is not only a model but a set of models that capture the structure of a system by using different types of modeling paradigms such as continuous, discrete, agent-based, rule-based, geometric-based, and so on. It is a system's duplicate or 'twin' to enable simulations and predict consequences, including during environmental changes.

The metaverse can be the backbone of creating Digital Twins of leaders. Then the leader can play with the virtual environment, get trained for situations, and predict situations of specific decisions that can go from operating decisions to utilizing/improving emotional intelligence. This idea we will see it in the next decades. This fusion of the metaverse with the Digital Twin will create the ultimate development platform.

References

Allen, C. (2020). The potential health benefits of meditation. *ACSM's Health & Fitness Journal, 24*(6), 28–32. https://doi.org/10.1249/FIT.0000000000000624.

Allison, C., Campbell, A., Davies, C., Dow, L., Kennedy, S., Mccaffery, J., Miller, A., Oliver, I., & Perera, I. (2012). Growing the use of virtual worlds in education: An OpenSim perspective. *Proceedings of the 2nd European Immersive Education Summit*, Paris (France), 26–27 November 2012.

Almalki, H. (2016). *A holistic framework for effective engineering leadership development using 3D virtual world simulation*. The University of Central Florida. Electronic Theses and Dissertations, 5091. Retrieved from https://stars.library.ucf.edu/etd/5091.

Bamodu, O., & Ye, X. M. (2013, September). Virtual reality and virtual reality system components. *Advanced Materials Research, 765–767*, 1169–1172. https://doi.org/10.4028/www.scientific.net/amr.765-767.1169.

Bartlett, C., Hall, B., & Bennett, N. (2007, June). *GE's Imagination breakthroughs: The Evo project*. Harvard Business School Case, 907-048. Retrieved from https://www.hbs.edu/faculty/Pages/item.aspx?num=34629.

Bennis, W., & O'Toole, J. (2000, May–June). Don't hire the wrong CEO. *Harvard Business Review, 78*(3), 170–176, 218. PMID: 11183978.

Bhide, S., Rabelo, L., Pastrana, J., Katsarsky, A., & Ford, C., (2015). *Development of virtual reality environment for safety training*. IISE Annual Conference Proceedings, 2302–2312. Retrieved from https://search.proquest.com/docview/1792029019?accountid=10003.

Birk, M. (2020, March 22). Why leaders need meditation now more than ever. *Harvard Business Review*. Retrieved from https://hbr.org/2020/03/why-leaders-need-meditation-now-more-than-ever.

Birshan, M., Meakin, T., & Strovink, K. (2107, April). What makes a CEO exceptional? *McKinsey Quarterly*. McKinsey & Company.

Brant, J., Dooley, R., & Iman, S. (2008). Leadership succession: An approach to filling the pipeline. *Strategic HR Review, 7*, 17–24.

Briggs-Myers, I., McCaulley, M., Quenk, N. L., & Hammer, A. L. (1998). *Manual: A guide to the development and use of the Myers-Briggs Type indicator* [measurement instrument]. Palo Alto, CA: Consulting Psychologists Press.

Brynjolfsson, E., Renshaw, A. A., & Van Alstyne, M. (1997). The matrix of change. *Sloan Management Review, 38*(2), 37–54. Retrieved from https://www.proquest.com/scholarly-journals/matrix-change/docview/224966246/se-2.

Çapin, T. K., Pandzic, I. S., Magnenat-Thalmann, N., & Thalmann, D. (1999, July 30). *Avatars in networked virtual environments* (1st ed.). Hoboken, NJ: Wiley.

Chorafas, D. N., & Steinmann, H. (1995). *Virtual reality: Practical applications in business and industry*. Englewood Cliffs, NJ: Prentice Hall.

Cruz-Benito, J., Therón, R., García-Peñalvo, F. J., & Pizarro Lucas, E. (2015, June). Discovering usage behaviors and engagement in an educational virtual world. *Computers in Human Behavior, 47*, 18–25. https://doi.org/10.1016/j.chb.2014.11.028.

Davis, C. (2020). *Investigation in engineering leadership using system engineering and virtual reality.* The University of Central Florida. Electronic Theses and Dissertations, 204. Retrieved from https://stars.library.ucf.edu/etd2020/204.

Efrati, A., & Bensinger, G. (2012, July 23). Ousted Yahoo chief lands new CEO role. *The Wall Street Journal.* Retrieved June 1, 2022, from https://www.wsj.com/art icles/SB10000872396390443570904577544813207778788.

Efrati, A., & Letzing, J. (2012, July 17). Google's Mayer takes over as Yahoo Chief. *The Wall Street Journal.* Retrieved June 5, 2022, from https://www.wsj.com/art icles/SB10001424052702303754904577531230541447956.

Elattar, A. (2014). *A holistic framework for transitional management.* The University of Central Florida. Electronic Theses and Dissertations, 4618. Retrieved from https://stars.library.ucf.edu/etd/4618.

Escobar, C., McGovern, M., & Morales-Menendez, R. (2021). Quality 4.0: A review of big data challenges in manufacturing. *Journal of Intelligent Manufacturing, 32*, 2319–2334. https://doi.org/10.1007/s10845-021-01765-4.

Girvan, C. (2018, February 8). What is a virtual world? Definition and classification. *Educational Technology Research and Development, 66*(5), 1087–1100. https://doi.org/10.1007/s11423-018-9577-y.

Gittleson, K. (2012, January 19). Can a company live forever? *BBC.* Retrieved from http://www.bbc.co.uk/news/business-16611040.

Goldman, D. (2011, December 30). Mark Hurd's sex scandal letter emerges. *CNN Money.* Retrieved from http://money.cnn.com/2011/12/30/technology/hurd_letter/index.htm.

Groth, A., & Bhasin, K. (2011, July 08). The best CEOs of the past 20 years. *Business Insider.*

Hauser, J., & Clausing, D. (1988). The house of quality. *Harvard Business Review,* May–June, No. 3, 63–70. Retrieved from https://hbr.org/1988/05/the-house-of-quality.

Hudson, K., Taylor, L. A., Kozachik, S. L., Shaefer, S. J., & Wilson, M. L. (2014, November 25). Second life simulation as a strategy to enhance decision-making in diabetes care: A case study. *Journal of Clinical Nursing, 24*(5–6), 797–804. https://doi.org/10.1111/jocn.12709.

Koploy, M. (2012, February 15). The 5 best (and worst) tech executives of all time. Retrieved from http://blog.softwareadvice.com/articles/enterprise/best-and-worst-tech-execs-1021511/.

Kouzes, J., & Posner, B. (2007). *The leadership challenge* (4th ed.). San Francisco, CA: John Wiley & Sons, Inc.

Laval, S., Gagan, A., Kamarsu, P., & Hears, M. (2012). Final Project F2/EIN5108_CV83. Dr. Luis Rabelo's Course EIN5108 (The Environment of Technical Organizations). The University of Central Florida.

Lemheney, A. J., Bond, W. F., Padon, J. C., LeClair, M. W., Miller, J. N., & Susko, M. T. (2016). Developing virtual reality simulations for office-based medical emergencies. *Journal of Virtual Worlds Research, 9*(1), 1–18.

Liberati, A., Altman, D., Tetzlaff, J., Mulrow, C., Gøtzsche, P., Ioannidis, J., Clarke, M., Devereaux, P., Kleijnen, J., & Moher, D. (2009). The PRISMA statement for reporting systematic reviews and meta-analyses of studies that evaluate health care interventions: Explanation and elaboration. *PLoS Medicine*, *6*(7), e1000100. https://doi.org/10.1371/journal.pmed.1000100. Epub 2009 July 21. PMID: 19621070; PMCID: PMC2707010.

Martin, C. (2012, March 02). Does the difficult boss always deserve to be fired? *Forbes*, Retrieved from http://www.forbes.com/sites/work-in-progress/2012/03/02/does-the- difficult-boss-always-deserve-to-be-fired/.

McKinsey & Co. (2022). *Value creation in the metaverse*. [online] McKinsey & Company. Retrieved from https://www.mckinsey.com/capabilities/growth-marketing-and-sales/our-insights/value-creation-in-the-metaverse.

Millikin, J., & Fu, D. (2005). The global leadership of Carlos Ghosn at Nissan. *Thunderbird International Business Review*, *47*(1), 121–137.

Naquin, C. E., & Kurtzberg, T. R. (2017). Leadership selection and cooperative behavior in social dilemmas: An empirical exploration of assigned versus group-chosen leadership. *Negotiation and Conflict Management Research*, *11*(1), 29–52. https://doi.org/10.1111/ncmr.12114.

Navarro-Haro, M. V., López-del-Hoyo, Y., Campos, D., Linehan, M. M., Hoffman, H. G., García-Palacios, A., et al. (2017). Meditation experts try virtual reality mindfulness: A pilot study evaluating the feasibility and acceptability of virtual reality to facilitate mindfulness practice in people attending a mindfulness conference. *PLoS One*, *12*(11), e0187777. https://doi.org/10.1371/journal.pone.0187777.

Pellerin, C. (2009). *How NASA builds teams: Mission critical soft skills for scientists, engineers, and project teams*. Hoboken, NJ: John Wiley & Sons.

Rajoria, P., Sharma, A., Sharma, M., & Sumaiya, B. (2022). Leadership style and organisational success. *World Journal of English Language*, *12*(3), 71.

Sequeira, L., & Morgado, L. (2013, March 1). Mechanisms of three-dimensional content transfer between the OpenSimulator and the Second Life Grid® platforms. *Journal of Gaming & Virtual Worlds*, *5*(1), 41–57. https://doi.org/10.1386/jgvw.5.1.41_1.

Siewiorek, A., Saarinen, E., Lainema, T., & Lehtinen, E. (2012, January). Learning leadership skills in a simulated business environment. *Computers & Education*, *58*(1), 121–135. https://doi.org/10.1016/j.compedu.2011.08.016.

Sorcher, M., & Brant, J. (2002). Are you picking the right leaders? *Harvard Business Review*, *80*, 78–85, 128.

Spain, E. (2020). Reinventing the leader-selection process: The US Army's new approach to managing talent. *Harvard Business Review*, *98*(6), 78–85.

Spector, M., & Mattioli, D. (2012, January 20). Can bankruptcy filing save Kodak?—Doubts persist on printer, patent strategy as icon seeks chapter 11 protection. *Wall Street Journal*, pp. 1–B.1.

Stewart, T. (2006). Growth as a process: An interview with Jeffrey R. Immelt. *Harvard Business Review: HBR*. Boston, MA: Harvard Business School Publ. Corp, ISSN 0017-8012, ZDB-ID 23826. pp. 60–71. Retrieved from http://www.ge.com/files/usa/stories/en/Growth_The HBR_Interview.pdf

Thoroughgood, C., Sawyer, K., Padilla, A., & Lunsford, L. (2018). Destructive leadership: A critique of leader-centric perspectives and toward a more holistic definition. *Journal of Business Ethics*, *151*(3), 627–649.

Usmani, F. (2022, March 24). Risk assessment matrix: Definition, examples, and templates. *PM Study Circle*. Retrieved from https://pmstudycircle.com/risk-assessment-matrix/.

Vinkenburg, C., Jansen, P., Dries, N., & Pepermans, R. (2014). Arena: A critical conceptual framework of top management selection. *Group & Organization Management, 39*, 33–68.

Williams-Bell, F. M., Kapralos, B., Hogue, A., Murphy, B. M., & Weckman, E. J. (2014, May 7). Using serious games and virtual simulation for training in the fire service: A review. *Fire Technology, 51*(3), 553–584. https://doi.org/10.1007/s10694-014-0398-1.

Yurieff, K. (2018, February 23). Snapchat stock loses $1.3 billion after Kylie Jenner tweet. *CNN.* Retrieved from https://money.cnn.com/2018/02/22/technology/-snapchat-update-kylie-jenner/index.html.

Printed in the United States
by Baker & Taylor Publisher Services